SPSS Manual

for

Moore and McCabe's

Introduction to the Practice of Statistics

FIFTH EDITION

LINDA SORENSEN
Algoma University

W. H. Freeman and Company
New York

ISBN: 0-7167-6363-X

EAN: 9780716763635

Printed in the United States of America

First printing

W. H. Freeman and Company
41 Madison Avenue
New York, NY 10010
Houndmills, Basingstoke RG21 6XS, England

www.whfreeman.com

Table Of Contents

Preface

For me, the use of software makes statistics very interesting. Tasks that were once too tedious for me to even contemplate are now readily completed with software such as SPSS. Advances in statistical software have opened up a whole new frontier for exploration. I hope that you will come to share my enthusiasm for this exploration as you work your way through this course.

This manual is a supplement to the fifth edition of Moore and McCabe's *Introduction To The Practice of Statistics* (IPS). The purpose of this manual is to show students how to perform the statistical procedures discussed in IPS using SPSS 12.0 for Windows. This manual provides applications and examples for each chapter of the text. The process was guided by the principles put forth by the American Statistical Association for teaching statistics. Most of the examples and subsequent discussion come directly from IPS. Step-by-step instructions describing how to carry out statistical analyses using SPSS 12.0 for Windows are provided for each topic covered in the text.

SPSS has been regarded as one of the most powerful statistical packages for many years. It performs a wide variety of statistical techniques ranging from descriptive statistics to complex multivariate procedures. In addition, a number of improvements have been made to version 12.0 of SPSS for Windows that make it more user-friendly.

Writing this manual has been a learning experience for me. I would like to think that I've been a good learner and that this will do two things: I believe that it will improve my teaching, and I believe that it will improve your learning.

I would like to thank Amy Shaffer, Laura Hanrahan, and Christopher Spavins of W. H. Freeman and Company, for assisting me with this project. Also, I truly appreciate the able assistance of Sarah Campbell.

I dedicate this work to my mother, Ida Marie Hunter (1924-2004): for the Lambert in me!

Chapter 0. Introduction to SPSS 12.0 for Windows

Topics covered in this chapter:

0.1 Accessing SPSS 12.0 For Windows
0.2 Entering Data
0.3 Saving An SPSS Data File
0.4 Opening An Existing SPSS Data File From Disk
0.5 Opening An SPSS Data File From The W. H. Freeman Web site
0.6 Opening A Microsoft Excel Data File From Disk
0.7 Defining A Variable
0.8 Recoding A Variable
0.9 Deleting A Case From An SPSS Data File
0.10 Opening Data Sets Not Created By SPSS Or Windows Excel
0.11 Printing In SPSS For Windows
0.12 Copying From SPSS Into Microsoft Word 97
0.13 Using SPSS Help

This manual is a supplement to *Introduction to the Practice of Statistics*, Fifth Edition, by David S. Moore and George P. McCabe, which is referred to as IPS throughout the manual. The purpose of this manual is to show students how to perform the statistical procedures discussed in IPS using SPSS 12.0 for Windows. This manual is not meant to be a comprehensive guide to all procedures available in SPSS for Windows. The instructions included here will work for most versions of SPSS for Windows.

Throughout this manual, the following conventions are used: (1) variable names are given in boldface italics (e.g., *age*); (2) commands you click or text you type are boldface (e.g., click **Analyze**); (3) important statistical terms are boldface; (4) the names of boxes or areas within an SPSS window are in double quotes (e.g., the "Variable Name" box); and (5) in an example number, the digit(s) before the decimal place is the chapter number and the digit(s) after the decimal place is the example number within that chapter (e.g., Example 1.3 is the third example in Chapter 1). Unless otherwise specified, all example, table, and figure numbers refer to examples, tables, and figures within this manual.

This chapter serves as a brief overview of tasks such as entering data, reading in data, saving data, printing output, and using SPSS Help that will help you get started in SPSS for Windows.

0.1 Accessing SPSS 12.0 For Windows

Find out how to access SPSS at your location. Figure 0.1 shows the opening screen in SPSS 12.0 after the software has been activated. Note that in the section labelled "Open an existing file" located within the "SPSS for Windows" window, the most recent files used in SPSS are listed, and you can open them easily by clicking on the desired file name and then clicking " OK."   If you want to enter data into a new file, you can click on "Type in data" and then click "OK." For new SPSS users, I recommend clicking on "Run the tutorial" to become acquainted with the software package.

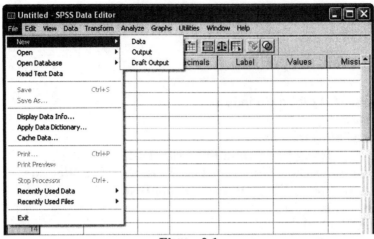

Figure 0.1

The window in the background of Figure 0.1 is the SPSS Data Editor. You will note that it is in a spreadsheet format, where columns represent variables and rows represent individual cases or observations. The SPSS Menu bar (**File, Edit, View…, and Help**) appears directly below "Untitled – SPSS Data Editor." Each of these main menu options contains a submenu of additional options. Throughout this manual, the first step in performing a particular task or analysis typically gives the SPSS Menu bar as the starting point (e.g., Click **File**, then click **Open**).

In addition to the Data Editor, the other primary window is the Output – SPSS Viewer window (which is not accessible until output has been generated). To move between these two windows, select **Window** from the SPSS Main Menu and click on the name of the desired window. In Figure 0.2, the SPSS Data Editor is the current active window, as shown by the ✔ in front of that window name.

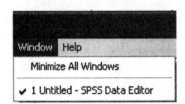

Figure 0.2

Most of the output in SPSS can be generated by clicking on a series of commands using a sequence of pulldown menus. However, it is also possible to generate output by writing a program using SPSS syntax language. The explanation of the syntax language is beyond the scope of this manual.

0.2 Entering Data

Before entering a data set into SPSS, determine whether the variable is quantitative or categorical. A **categorical** variable places an individual into one of several groups or categories, and a **quantitative** variable assumes numerical values for which arithmetic operations make sense. SPSS further classifies variables as ordinal, nominal, or scale. An **ordinal** variable classifies characteristics about the objects under study into categories that can be logically ordered. Some examples of ordinal variables are the size of an egg (small, medium, or large) and class standing (freshman, sophomore, junior, or senior). A **nominal** variable classifies characteristics about the objects under study into categories. Some examples of nominal variables are eye color, race, and gender. **Scale** variables collectively refer to both interval and ratio variables and are quantitative variables for which arithmetic operations make sense. Some examples of scale variables are height, weight, and age. By default, SPSS assumes that any new variable is a scale variable formatted to have a width of 8 digits, 2 of which are decimal places. This is denoted in SPSS by "numeric 8.2."

Suppose you want to enter the data set presented in Example 1.2 in IPS into SPSS for Windows. The variables in this data set include one categorical and two quantitative variables.

To create this data set in SPSS for Windows, follow these steps:

1. Click **File,** click **New,** and then click **Data.** The SPSS Data Editor is cleared.
2. Enter the data from Example 1.2 (see Figure 0.3).

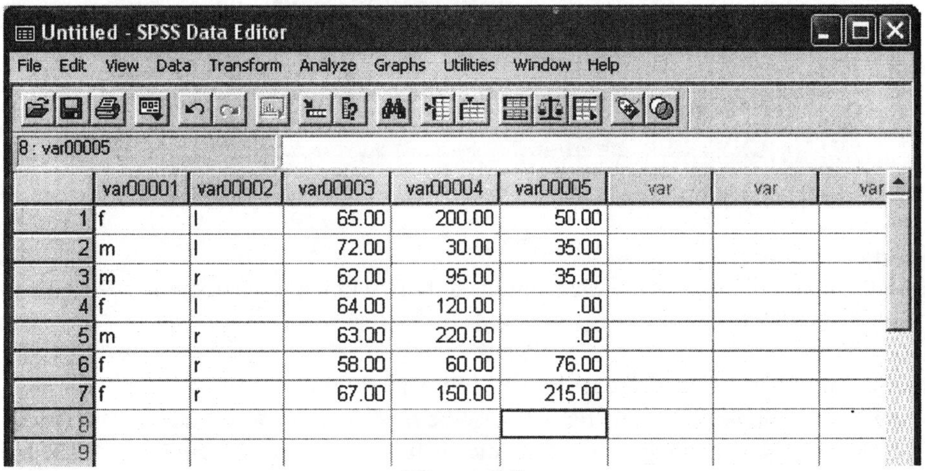

	var00001	var00002	var00003	var00004	var00005	var	var	var
1	f	l	65.00	200.00	50.00			
2	m	l	72.00	30.00	35.00			
3	m	r	62.00	95.00	35.00			
4	f	l	64.00	120.00	.00			
5	m	r	63.00	220.00	.00			
6	f	r	58.00	60.00	76.00			
7	f	r	67.00	150.00	215.00			
8								
9								

Figure 0.3

3. To define the variables, click on the **Variable View** tab at the bottom of the screen and make the appropriate changes. Note that SPSS has defaulted to the variable type, width and number of decimal places. You may change any of these if you wish. Note also that you can add a label, variable levels, and specify a code for missing values.
4. To change the variable name from the default name of ***var00001*** to a more appropriate name (such as ***sex***), type ***sex*** into the "Name" box. Variable names in SPSS can be at most 64 characters long, containing no embedded blanks. See Figure 0.4.

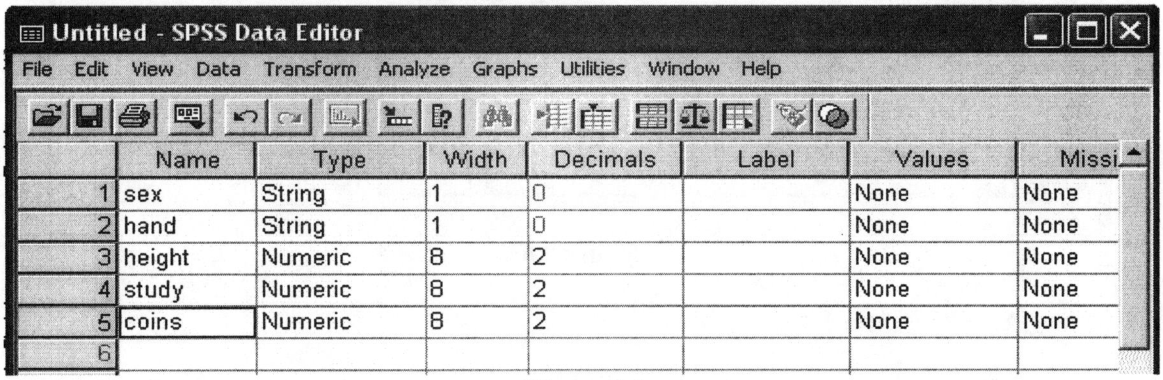

	Name	Type	Width	Decimals	Label	Values	Missi
1	sex	String	1	0		None	None
2	hand	String	1	0		None	None
3	height	Numeric	8	2		None	None
4	study	Numeric	8	2		None	None
5	coins	Numeric	8	2		None	None
6							

Figure 0.4

5. To change the type of the variable, click in the **Type** box and then click on the three dots that appear. Click **String** for a categorical variable. The "Variable Type" window shown in Figure 0.5 appears.

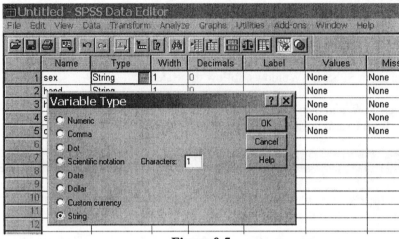

Figure 0.5

6. Click **String**.
7. To change the width of the variable *sex* from the default width of 8 characters to 9 characters, type **9** in the "Width" box.
8. Click **OK.**
9. The variable name *sex* will now appear in the SPSS Data Editor.
10. Each of the other variable characteristics can also be modified as appropriate. For example, you may wish to add value labels that indicate that 'f ' means female and 'm' a male subject (see Figure 0.6). Enter the value, the value label and then click **ADD**. When you are finished adding labels for each of the values of a variable, click **OK**.

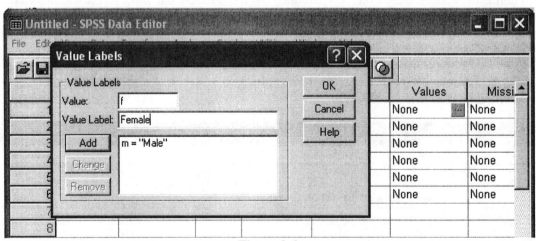

Figure 0.6

0.3 Saving An SPSS Data File

To save an SPSS data file, follow these steps:

1. Click **File** and click **Save as.** The "Save Data As" window appears; see Figure 0.7.

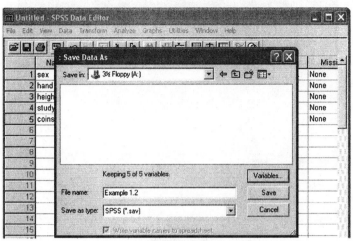

Figure 0.7

2. If you wish to save the file on disk, click ▼ in the "Save in" box until **3½ Floppy [A:]** appears, and then click on this destination. If you prefer to save the file in a different location, continue to click on ▼ until the name of the desired location is found, then click on this location name.
3. By default, SPSS assigns the .sav extension to data files. If you wish to save the data in a format other than an SPSS data file, click ▼ in the "Save as type" box until the name of the desired file type appears, then click on this file type name.
4. In the "File name" box, type in the desired name of the file you are saving. The name **Example 1.2** is used in Figure 0.7. Make sure a disk is in the A: drive, and then click **Save.**

0.4 Opening An Existing SPSS Data File From Disk

To open an existing SPSS data file from a disk, follow these steps:

1. Click **File** and click **Open.** The "Open File" window appears (see Figure 0.8).
2. If you wish to open an SPSS data file from a disk in the A: drive, click ▼ in the "Look in" box until **3½ Floppy [A:]** appears, then click on this name. If you prefer to open a file stored in a different location, continue to click on ▼ until the name of the desired location is found, then click on this location name.
3. All files with a ".sav" extension will be listed in the window. For other types of files click on "All files:" so that other file types appear in the window as well. Click on the name of the data file you wish to open. This name now appears in the "File name" box. In Figure 0.8, *Example 1.2.sav* is the desired SPSS data file.
4. Click **Open.**

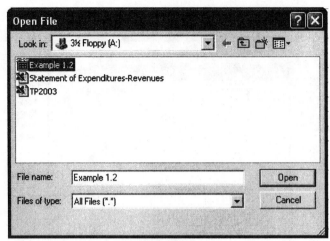

Figure 0.8

0.5 Opening An SPSS Data File From The W. H. Freeman Web site

To open an SPSS Data File from the W. H. Freeman Web site for IPS, go to www.whfreeman.com/ips5e and look for the appropriate button (something like SPSS data sets). At the time of writing this manual, final details are not yet available. There are additional tools including SPSS macros and sample quizzes at this site. I recommend that you visit it often.

0.6 Opening A Microsoft Excel Data File From Disk

To open a Microsoft Excel data file in SPSS for Windows, follow these steps:

1. From the SPSS Data Editor Menu bar, click **File** and then click **Open.** The SPSS "Open File" window appears as it did in Figure 0.8.
2. If the data file was saved in a different location, continue to click on ▼ until the name of the appropriate location appears, and then click on this location name.
3. Click on the name of the file you wish to open in SPSS for Windows. The desired file name will appear in the "File name" box. In this example, as shown in Figure 0.9, the name of the desired Microsoft Excel file is *US General Social Survey 1991.xcl.* SPSS will then convert the Excel file to an SPSS file. Follow the instructions given on-screen if you need to open a file that is not in SPSS or Excel format, for example ".por" or ".sav" format.

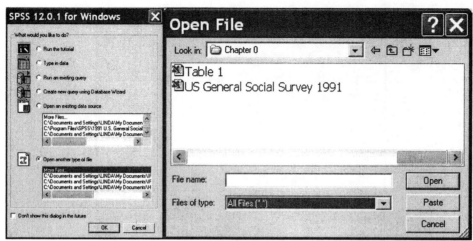

Figure 0.9

4. Click **Open.** The "Opening Excel Data Source" window appears, as shown in Figure 0.10. If the data set contained the variable names in the first row, click on the "Read variable names" box so that a check mark appears in the box. For this example, the variable names appear in the first row of the data set (see Figure 0.11); therefore, this box should be checked.

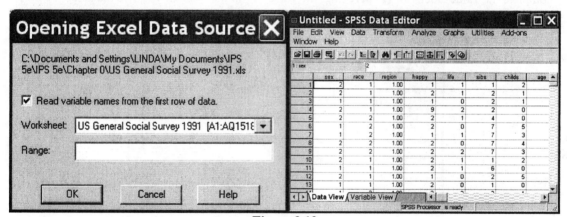

Figure 0.10

5. Click **OK.**
6. The data appear in the "Untitled – SPSS Data Editor" window as shown to the right in Figure 0.10.

0.7 Defining A Variable

You might be interested in changing the default SPSS variable names to more appropriate variable names. **Variable names** in SPSS can be at most 8 characters long, containing no embedded blanks. However, you can add a **variable label,** a descriptive explanation of the variable name. Variable labels can be at most 256 characters long and can contain embedded blanks. The use of the variable label can make the SPSS output more informative.

The example used in this section is a small sample based on real data for individuals who completed the Social Problem Solving Inventory – Short Form (SPSI-SF). See Figure 0.11 for the data. The data have already been entered in SPSS for Windows.

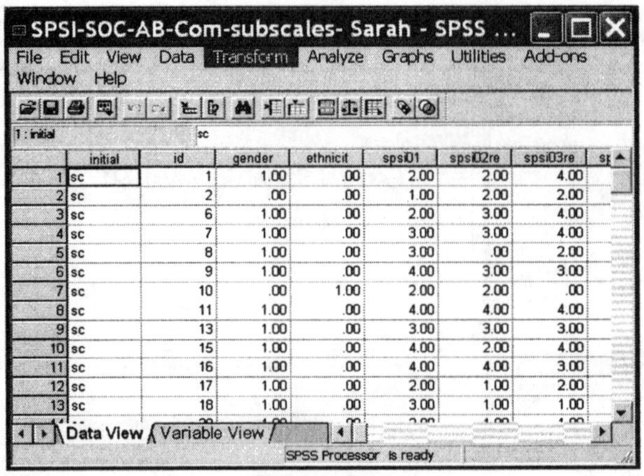

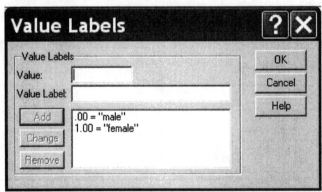

Figure 0.11

To define a variable in SPSS (such as declaring the variable name, declaring the variable type, adding variable and value labels, and declaring the measurement type), follow these steps:

1. Click on the "Variable View" tab at the bottom left of the data set.   Figure 0.12 shows the process of adding the labels "Male" and "Female" to **gender**. The value "0 = Male" has already been added. To complete the value labels for **gender,** enter the value "1" in the "Value" box and the label "Female" in the "Value Label" box and then click **Add,** then **OK.**

Figure 0.12

2. The value labels also have been added for **spsi01** and they are shown in Figure 0.13.

Figure 0.13

3. Select the appropriate measurement from the options available in the "Measurement" box: scale, ordinal, or nominal. The default option for numeric variables is scale.
4. If you are interested in adding additional information about the characteristics of your variables, click in the appropriate box and follow the instructions on the screen.

0.8 Recoding A Variable

Some of the analyses to be performed in SPSS will require that a categorical variable be entered into SPSS as a numeric variable (e.g., *gender* from the previous illustration). If the variable has already been entered as a string into the SPSS Data Editor, you can easily create a new variable that contains the same information as the string variable but is numeric simply by recoding the variable. In the example that follows, I have changed the string value for *initial* from "sc" to "1". When you recode variables like this, use numbers or string variables that make sense for your data.

To recode a variable into a different variable (the recommended option), follow these steps:

1. Click **Transform,** click **Recode,** and then click **Into Different Variables.** The "Recode into Different Variables" window in Figure 0.14 appears. Enter a new variable name (no more than 8 characters) in the "Name" box. In this example, I have used *init rec* as my new variable name with a label of *initial recoded*.

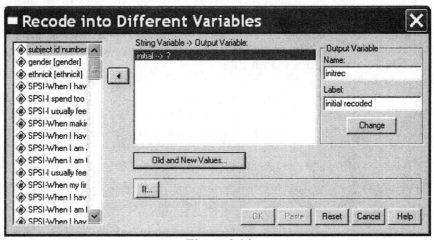

Figure 0.14

2. Click **Old and New Values.** The "Recode into Different Variables: Old and New Values" window, also shown in Figure 0.15, appears.

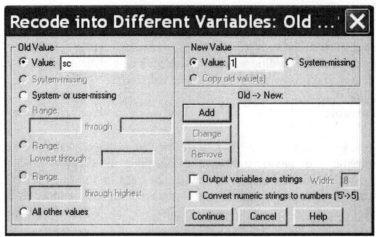

Figure 0.15

3. In the "Value" box within the "Old Value" box, type **sc** (the value has to be entered into this box *exactly* as it appears in the SPSS Data Editor). In the "Value" box in the "New Value" box, type **1** (or whatever you wish to change it to). Click **Add.** "sc→1" appears in the "Old→ New" box. If you have more than 1 level of your variable, continue with this process until all have been assigned a unique numerical (or string if that is more appropriate) value. Remember that each time the value is entered into the "Old Value" box it must appear *exactly* as it appears in the SPSS Data Editor. When all values have been recoded, click **Continue.**
4. Click **OK.** The new variable *sys rec* appears in the SPSS Data Editor.

0.9 Deleting A Case From An SPSS Data File

Occasionally, extreme values (outliers) will be contained in data sets. It may be desirable to generate analyses both with and without the outlier(s) to assess the effect it (they) may have. Visual displays such as Stemplots, Histograms, and Bar Graphs (see Chapter 1) help you to understand your data and determine if a case should be deleted from your analysis. In what follows, I have assumed that you already know the number of the case you want deleted. To remove case 24 from the SPSS data file, follow these steps:

1. Use the scroll bar along the left side of the SPSS Data Editor so the desired case number appears among the numbers in the left-most, gray-colored column. Click on the desired case number. The entire row associated with that case number is highlighted (see Figure 0.16).
2. Click **Edit** and then click **Clear.** The row for that case number will be deleted. NOTE: Unless you want this case to be permanently deleted, **DO NOT SAVE THE DATA SET** upon exiting SPSS or the information for this case will be lost.

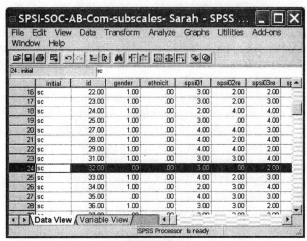

Figure 0.16

3. *NOTE:* With large data sets, you can also go to the desired case number by clicking **Data** and then clicking **Go to case.** The "Go to Case" window appears (see Figure 0.17). Type in the desired case number, click **OK,** and then click **Close.** The first cell of that case number will be highlighted. You will then need to click on the case number in the left-most, gray-colored column.

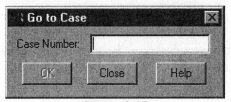

Figure 0.17

These steps can be repeated if you want to delete other cases. *CAUTION:* Once an observation is deleted, SPSS renumbers all subsequent cases. Therefore, the case number for a subsequent case will change once an earlier case number has been deleted.

0.10 Opening Data Sets Not Created By SPSS Or Windows Excel

Although many types of data files can be read directly into SPSS for Windows, I recommend initially reading the data file into Microsoft Excel and then reading the Microsoft Excel file into SPSS for Windows as shown above. In this example, I will import data in a .dat format. This type of data may have been produced with a word processor such as Word for Windows.

To convert a text data file into a Microsoft Excel file, follow these steps:

1. Open Microsoft Excel.
2. Click **File** and then click **Open.** The "Open" window appears.
3. Click ▼ in the "Look in" box until you find the name of the drive that contains the text data file, and then click on this name.
4. Click on the file name for the file that you wish to open. The "Text Import Wizard" window opens. The "Text Import Wizard" first determines whether the data are of fixed width (variable values are lined up in columns). The user can set the "Text Import Wizard" to begin importing the data at row 2, which is correct if variable names appear as column headings in the ".dat" file.

5. Click **Next.** The second window of the "Text Import Wizard" appears. Each column of numbers in the "Data preview" box represents the different variables in the data set. The "Text Import Wizard" suggests where the column breaks should occur between the variables. Additional column breaks can be inserted by clicking at the desired location within the "Data preview" box. An existing column break can be removed by double-clicking on that column break.

6. Click **Next.** The third window of the "Text Import Wizard" appears. Here you can exclude variables, if desired, by clicking on the column you do not wish to import and selecting "Do not import column (Skip)" in the "Column data format" box. You can also specify the format for each variable. By default, all variables are assigned a General format; this provides the most flexibility in SPSS for Windows.

7. Click **Finish.** The data will appear in the Microsoft Excel spreadsheet, where the columns represent variables and the rows represent cases.

8. Click "Save As" and give your Excel file a name.

9. Now use SPSS for Windows to open the Excel file.

0.11 Printing In SPSS For Windows

Printing Data

To print out the data in the SPSS Data Editor, follow these steps:

1. Make the SPSS Data Editor the active window (click somewhere on the SPSS Data Editor).
2. Click **File,** then click **Print.** A "Print" window similar to the one shown in Figure 0.18 will appear.
3. SPSS assumes that you want to print the entire data set. If you wish to print only a subset of the data (e.g., only one variable), you must first click on the appropriate column heading(s). If this is done, "Selection" in Figure 0.18 rather than "All" will be highlighted.
4. Click **OK.**

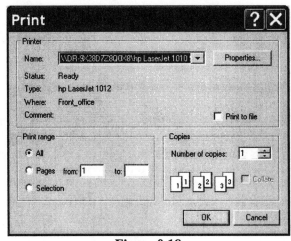

Figure 0.18

Printing Output And Charts In SPSS For Windows

To print out SPSS output and SPSS charts, follow these steps:

1. Make the "Output – SPSS Viewer" window the active window.

2. If you want to print all of the output, including graphs, click the Output icon (⊞ Output), which appears at the top of the left-hand window. All output is highlighted, as shown in Figure 0.19.

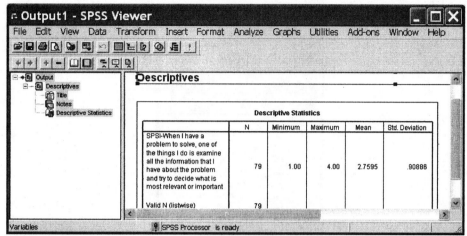

Figure 0.19

3. If you want to print only a specific portion of the output, such as only the Descriptive Statistics, click on that icon (see Figure 0.20).
4. Click **File,** click **Print,** and then click **OK.**
5. After the output has been printed, it can be deleted. This is accomplished by first highlighting the portion of the output you wish to delete. For example, you can highlight all of the output, as shown in Figure 0.19, or highlight a specific icon, as shown in Figure 0.20. Then after the desired output has been selected, you can delete the output by pressing the **Delete** key.

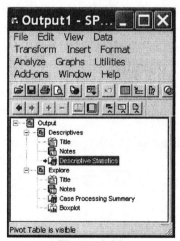

Figure 0.20

0.12 Copying From SPSS Into Microsoft Word 97

Copying A Chart Or Table

To copy a chart or table from SPSS into Microsoft Word 97, follow these steps (note: It is assumed that Microsoft Word 97 has already been opened):

1. In the "Output – SPSS Viewer" window, select the chart or table to be copied by clicking on the icon that appears in the left side of the window (for example, the **Boxplot** icon shown in Figure 0.21).

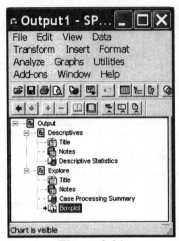

Figure 0.21

2. Click **Edit** and then click **Copy objects.**
3. In Microsoft Word 97, position the cursor at the desired place in the document, click **Edit,** and then click **Paste.**
4. From the Microsoft Word Main Menu, click **Format,** and then click **Picture.**
5. A "Format Picture" window appears. Click on the **Position** tab.
6. Click **Float over text**. The ✔ in front of "Float over text" disappears. For other versions of Microsoft Word, explore the advanced options.
7. Click **OK.**
8. If you are interested in resizing the chart, click on the chart and place the cursor on one of the little black squares ("handles") that appear along the box that now outlines the chart. Using the right mouse button, drag the cursor in or out depending on how you want to resize the chart. Using one of the four corner squares will resize the chart but maintain the scale of width to height. The two center squares along the sides will change the width of the chart but maintain the current height. The two center squares along the top and bottom will change the height of the chart but maintain the current width. Care should be taken in resizing a chart because it is easy to distort the chart's overall image and the information the chart is meant to relay.
9. If you have chosen a figure, it will appear in Microsoft Word like Figure 0.22.

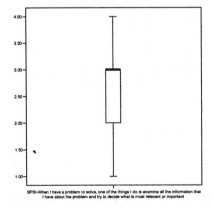

Figure 0.22

10. To begin typing in Microsoft Word 97, click on any area within the document that is outside the chart or table.

0.13 Using SPSS Help

SPSS has extensive and useful on-line help. Suppose you want to know the steps needed to obtain a boxplot using SPSS and no manual is available. This information can be obtained using the help function of SPSS by following these steps:

1. Click **Help,** and then click **Topics.** The "Help Topics" window appears, as shown in Figure 0.23.
2. Click the **Index** tab and type **Box** in the "Type the first few letters of the word you're looking for" box. Choose "boxplots" from the list of available options and information about creating boxplots appears in the right hand window as shown above.
3. Do the same with other options listed under boxplots to get full instructions on the different types available and how to create them.
4. To print the directions, click **Print**.
5. To exit SPSS Help, click ☒ in the upper right corner of the "SPSS for Windows" window (see Figure 0.23).

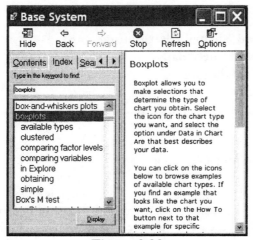

Figure 0.23

Postscript

I recommend that you develop a record-keeping system so that you will be able to find the files that you create in the future. Create a table in one of your notebooks to record information such as file name and location, chapter and exercise number reference, and a brief description of the data that are in the file.

Chapter 1. Looking At Data —Distributions

Topics covered in this chapter:

1.1 Displaying Distributions With Graphs

Statistical tools and ideas help us examine data. This examination is called **exploratory data analysis**. This section introduces the notion of using graphical displays to perform exploratory data analysis. The graphical display used to summarize a single variable depends on the type of variable being studied (i.e., whether the variable is categorical or quantitative). For categorical variables, **pie charts** and **bar graphs** are used. For quantitative variables, **histograms, stemplots,** and **time plots** are ideal.

Bar Graphs

The data shown on Page 1-7 in IPS regarding Education will be used to illustrate pie charts and bar graphs. These data are categorical. To make your own example, open a blank SPSS for Windows Data Editor screen (refer to Chapter 0 for a review of how to do this) and enter the education data in a string variable labeled *educatio,* and the counts and percents in numerical variables called *count* and *percent.*

To create a bar graph for a categorical variable (such as *educatio*), follow these steps:

1. Click **Graphs** and then click **Bar.** The "Bar Charts" window in Figure 1.1 appears.

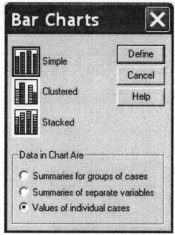

Figure 1.1

2. Click the button for "Data in Chart are Values of individual cases". Click on "Simple" then click **Define.** The "Define Simple Bar: Values of Individual Cases" window in Figure 1.2 appears.

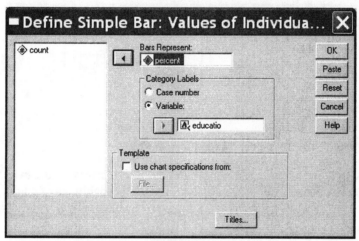

Figure 1.2

3. Click *educatio,* then click ▶ to move *educatio* into the "Variable" box.
4. To replicate Figure 1.1a of IPS, click *percent* and then click ▶ beside the "Bars Represent" box and *percent* will now appear in the box as in Figure 1.2 above.
5. If you are interested in including a title or a footnote on the chart, click **Titles.** The "Titles" window in Figure 1.3 on the following page appears. In the properly labeled box ("Title" or "Footnote"), type in the desired information.
6. Click **Continue.** Click **OK.**

Figure 1.4 on the following page is the resulting SPSS for Windows output.

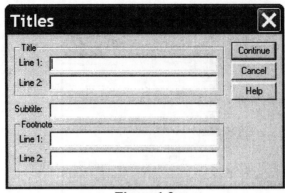

Figure 1.3

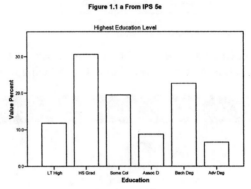

Figure 1.4

Editing Bar Graphs

Some instructions are included here for editing Bar graphs. These instructions can be used for editing other types of graphical output as well. A description of the full capabilities of editing charts using SPSS is beyond the scope of this manual. By experimenting, you will find that you have a wide range of options for your output.

There are several other options for bar graphs. For example, you may make other types of changes such as spacing the bars farther apart. See below for some of the steps involved.

1. Double-click on the bar graph in the "Output – SPSS for Windows Viewer" window. The bar graph now appears in the "Chart Editor" window, which has a new menu and tool bars. The chart in the "Output – SPSS for Windows Viewer" window will be shaded with black diagonal lines whenever the chart is in the "Chart Editor" window. See Figure 1.5 on the facing page.

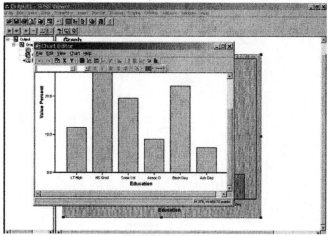

Figure 1.5

2. Double-click on one of the bars in the chart. The following option screens appear (Figure 1.6). Click on "Bar Options" and move the **Bars** slider to 50% to make the bars farther apart.
3. Click **Apply** and then click **Close.**

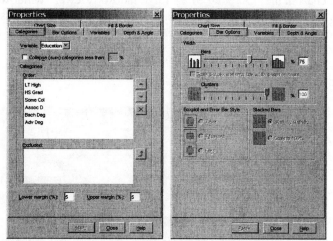

Figure 1.6

3. Click "Text" and then under "Style" change "normal" to "bold".
4. Click **Apply** and then **Close.**

Notice the difference between Figure 1.7 on the next page and Figure 1.4 produced earlier in this chapter.

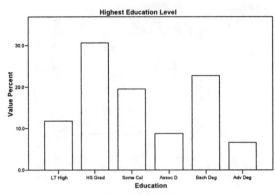

Figure 1.7

To change the color of an area within the chart (or to change the color of a pie slice), follow these steps:

1. The chart must be in the "Chart Editor" window. If the chart is not in the "Chart Editor" window, double-click on the chart in the "Output – SPSS for Windows Viewer" window.
2. Click on the area within the chart for which a change in color is desired (for instance, the bars in the bar graph). The area to be changed will be outlined in blue. For this example, the bars in the chart are outlined in blue.
3. Double-click one of the bars and the "Properties" window shown in Figure 1.8 appears. Click "Fill & Border" and a "window" like Figure 1.9 appears.

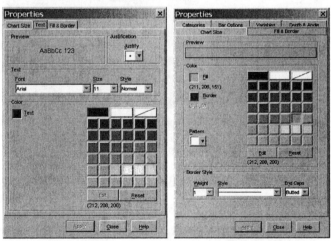

Figure 1.8   **Figure 1.9**

4. Make sure that "Fill" rather than "Border" is selected within the "Properties" box. Click on the desired color for the fill, for instance, light gray.
5. Click **Apply** and then click **Close.**
6. If you are finished editing the chart, click **File** from the main menu bar within the "Chart Editor" window and then click **Close** to return to the "Output – SPSS for Windows Viewer" window.

To change the pattern of an area within the chart, follow these steps:

1. The chart must be in the "Chart Editor" window. If the chart is not in the "Chart Editor" window, double-click on the chart in the "Output – SPSS for Windows Viewer" window.
2. Click on the area within the chart for which a change in pattern is desired (for instance, the bars in the bar graph). The area to be changed will be outlined in blue. For this example, the bars in the chart are outlined in blue.
3. Double-click one of the bars and the "Properties" window like Figure 1.7 appears. Click "Fill & Border" and then click "Pattern" and a variety of pattern options appear. See Figure 1.10.
4. Click the desired pattern. A solid color pattern is the default option.
5. Click **Apply** and then click **Close.**
6. If you are finished editing the chart, click **File** from the main menu bar within the "Chart Editor" window and then click **Close** to return to the "Output – SPSS for Windows Viewer" window.

Figure 1.10

To add data labels to the bars within the chart, follow these steps:

1. The chart must be in the "Chart Editor" window. If the chart is not in the "Chart Editor" window, double-click on the chart in the "Output – SPSS for Windows Viewer" window.
2. Click one of the bars in the chart. The area to be changed will be outlined in blue. For this example, the bars in the chart are outlined in blue.
3. Right-click one of the bars and click on "Show Data Labels". Next, a "Properties" window appears (see Figure 1.11 on the following page).
4. As a default, the middle option is chosen. To keep this option, simply click **Close** on the "Properties" window and the data labels will be in the middle of the bars in the chart, as shown in Figure 1.12.
5. To choose another position for the data labels, select one of the other options under the heading **Label Position.**
6. Click **Apply** and then click **Close.**
7. The data labels will appear in the chart according to the option selected.
8. If you are finished editing the chart, click **File** from the main menu bar within the "Chart Editor" window and then click **Close** to return to the "Output – SPSS for Windows Viewer" window.

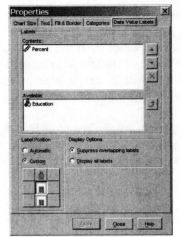

Figure 1.11

Figure 1.1 a From IPS 5e

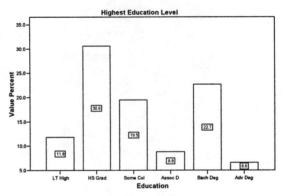

Figure 1.12

To make changes to the *x* axis (such as changing the axis title or the orientation of the labels), follow these steps:

1. The chart must be in the "Chart Editor" window. If the chart is not in the "Chart Editor" window, double-click on the chart in the "Output – SPSS for Windows Viewer" window.
2. Click the **X** icon on the toolbar at the top of the screen in the "Chart Editor" window. The "Properties" window in Figure 1.13 on the facing page appears.
3. To modify any of the text boxes (the *x* axis, for example), double-click on the text and make the desired changes. You could replace "Some Col" with "College". See Figure 1.14.
4. To change the title of the axis, double-click on the text and make the desired changes. You could replace "Education" with "Level of Education". See the right side of Figure 1.14.

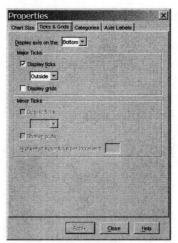

Figure 1.13

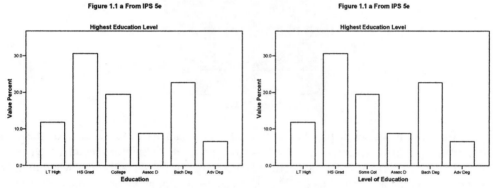

Figure 1.14

5. To change the orientation of the labels on the *x* axis, click ▼ in the "Label Orientation" box and then click the desired orientation style (for instance, **Diagonal**). See Figure 1.15.
6. Click **Apply** and then click **Close.**
7. If you are finished editing the chart, click **File** from the main menu bar within the "Chart Editor" window and then click **Close** to return to the "Output – SPSS for Windows Viewer" window.

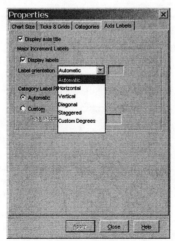

Figure 1.15

To make changes to the *y* axis (such as changing the axis title or the orientation of the labels), follow
these steps:

1. The chart must be in the "Chart Editor" window. If the chart is not in the "Chart Editor" window,
    double-click on the chart in the "Output – SPSS for Windows Viewer" window.
2. Click the **Y** icon on the toolbar at the top of the screen in the "Chart Editor" window. The
    "Properties" window in Figure 1.16 appears.

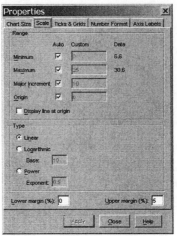

Figure 1.16

3. To change the range of values displayed on the *y* axis, for instance to 0 to 45, click on the **Scale** tab in
    the "Properties" window. Then uncheck the box (to the right of "Minimum") under the heading
    "Auto" and type **0** in the box under the heading "Custom". Do the same for the "Maximum" but type
    45 in the box under the heading "Custom". Using 0 as the minimum value for most charts is
    advisable to avoid generating misleading graphs.
4. To change the spacing of the tick marks on the *y* axis, for instance to 5, click on the **Ticks and Grids**
    tab in the "Properties" window. Then uncheck the box (to the right of "Major Increment") under the
    heading "Auto" and type **5** in the box under the heading "Custom". *NOTE*: The axis range must be
    an even multiple of the major increment. Figure 1.17 is the resulting SPSS for Windows Output.

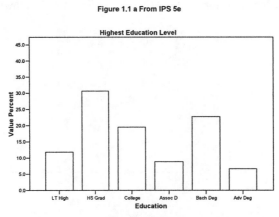

Figure 1.1 a From IPS 5e

Figure 1.17

4. Click **Apply** and then click **Close.**

5. If you are finished editing the chart, click **File** from the main menu bar within the "Chart Editor" window and then click **Close** to return to the "Output – SPSS for Windows Viewer" window.

Pie Charts

To create a pie chart for a categorical variable (such as *education* used in the bar graph examples above), follow these steps:

1. Click **Graphs** and then click **Pie.** The "Pie Charts" window in Figure 1.18 appears.

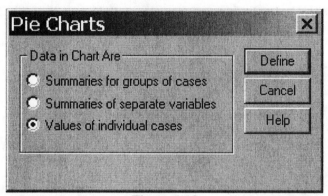

Figure 1.18

2. Click the button for "Data in Chart are Values of individual cases". Click on **Define.**
3. The "Define Pie: Values of Individual Cases" window in Figure 1.19 appears.
4. Click *educatio,* click "Variable" and then click ▶ to move *educatio* into the "Variable" box.
5. Next click *percent* and then click ▶ to move *percent* into the "Slices Represent" box.
6. If you are interested in including a title or a footnote on the chart, click **Titles.** The "Titles" window appears (see Figure 1.3). In the properly labeled box ("Title" or "Footnote"), type in the desired information.
7. Click **OK.**
8. To change the color or pattern-fill of the slices or to add data labels to the slices, follow the directions noted above in the Editing Bar Graphs section.

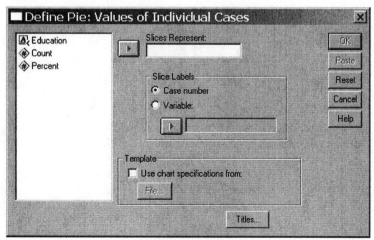

Figure 1.19

Additional Pie Chart Editing

To add additional data labels, hide the legend, or explode slices of pie charts, follow these steps:

1. The chart must be in the "Chart Editor" window. If the chart is not in the "Chart Editor" window, double-click on the chart in the "Output – SPSS for Windows Viewer" window.
2. Click any slice within the pie chart to outline the pie chart in blue.
3. Right-click any slice and click on "Show Data Labels". Next, a "Properties" window appears (see Figure 1.20).

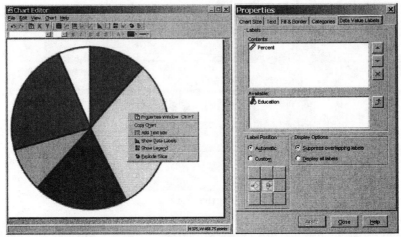

Figure 1.20

4. In the "Properties" window, click *Education* in the "Available box and then click ▶ to move *Education* into the "Contents" box.
5. Click **Apply** and then click **Close.**
6. Click on the legend to outline it in blue.
7. Right-click on the legend and then click "Hide Legend". Next, a "Properties" window appears.
8. Click **Close**.
9. Click any slice within the pie chart to outline the pie chart in blue.
10. Right-click any slice and click on "Explode Slice". Figure 1.21 is the resulting SPSS for Windows Output.

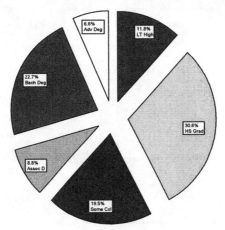

Figure 1.21

11. If you are finished editing the chart, click **File** from the main menu bar within the "Chart Editor" window and then click **Close** to return to the "Output – SPSS for Windows Viewer" window.

Histograms

A histogram breaks the range of values of a quantitative variable into intervals and displays only the count or the percentage of the observations that fall into each interval. You can choose any convenient number of intervals, but you should always choose intervals of equal width.

Table 1.1 in IPS presents service times in seconds for calls to a customer service center. The data can be retrieved from the Web site: www.whfreeman.com/ips5e. When you open the file, the SPSS for Windows Data Editor contains a variable called *servicet*.

To create a frequency histogram of this distribution, follow these steps:

1. Click **Graphs** and then click **Histogram.** The "Histogram" window in Figure 1.22 appears.
2. Click *percent,* then click ▶ to move *percent* into the "Variable" box.
3. If you want to include a title or a footnote on the chart, click **Titles.** The "Titles" window shown earlier in Figure 1.3 appears. In the properly labeled box ("Title" or "Footnote"), type in the desired information.

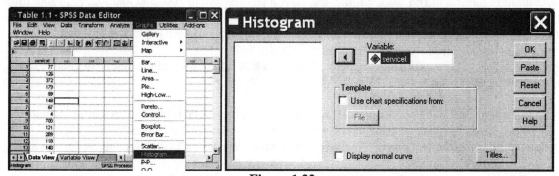

Figure 1.22

4. Click **Continue** and then click **OK.**

Figure 1.23 on the following page is the default histogram created by SPSS for Windows. The class intervals for the default histogram are 0 to under 500, 500 to under 1000, . . . , 2000 to under 2500.

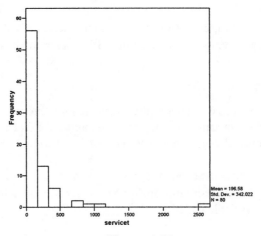

Figure 1.23

Editing Histograms

To obtain a picture of the distribution similar to that shown in Figure 1.2 of IPS, you can edit the histogram. A distinct difference between the SPSS for Windows default histogram and the histogram in Figure 1.2 of IPS is the chosen class intervals (and the fact that we have only 80 of the data points). The histogram in Figure 1.2 of IPS has the following class intervals: 0.0 to 100, 200 to 300, . . . , 1200-1300. To edit the histogram, double-click on the histogram in the "Output – SPSS for Windows Viewer" window to have it appear in the "Chart Editor" window, which has its own menu and tool bars.

To make changes to the x axis (such as changing the axis title, orientation of the labels, or change the class intervals), follow these steps:

1. The chart must be in the "Chart Editor" window. If the chart is not in the "Chart Editor" window, double-click on the chart in the "Output – SPSS for Windows Viewer" window.
2. To modify any of the text boxes (the title on the x axis, for example), double-click on the text and make the desired changes.
3. Click the **X** icon on the toolbar at the top of the screen in the "Chart Editor" window. The "Properties" window in Figure 1.24 on the facing page appears.
4. To change the orientation of the labels on the x axis, click the **Axis Labels** tab in the "Properties" window. Click ▼ in the "Label Orientation" box and then click the desired orientation style (see Figure 1.16 as an example).
5. Click **Apply** and then click **Close.**
6. To change the class intervals on the x axis, click the **Scale** tab in the "Properties" window.
7. Uncheck the ✔ beside "Major Increment" and enter the interval size you desire. For example, the current increment is set at 500; let's change it to 350. Figure 1.25 shows the results.
8. If you are finished editing the chart, click **File** from the main menu bar within the "Chart Editor" window and then click **Close** to return to the "Output – SPSS for Windows Viewer" window.

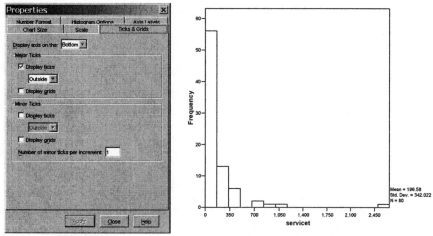

Figure 1.24   **Figure 1.25**

To make changes to the *y* axis (such as changing the axis title, orientation of the labels, or changing the class intervals), follow these steps:

1. The chart must be in the "Chart Editor" window. If the chart is not in the "Chart Editor" window, double-click on the chart in the "Output – SPSS for Windows Viewer" window.
2. To modify any of the text boxes (the title on the *y* axis, for example), double-click on the text and make the desired changes.
3. Click the **Y** icon on the toolbar at the top of the screen in the "Chart Editor" window. The "Properties" window in Figure 1.26 appears.

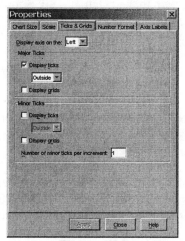

Figure 1.26

4. To change the orientation of the labels on the *y* axis, click the **Axis Labels** tab in the "Properties" window. Click ▼ in the "Label Orientation" box and then click the desired orientation style (see Figure 1.16 as an example).
5. Click **Apply** and then click **Close.**
6. To change the class intervals on the *x* axis, click the **Scale** tab in the "Properties" window.
7. Uncheck the ▾ beside "Major Increment" and enter the interval size you desire. For example, the current increment is set at 10; let's change it to 15. Figure 1.27 on the following page shows the results.

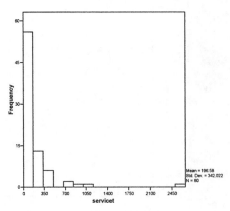

Figure 1.27

8. If you are finished editing the chart, click **File** from the main menu bar within the "Chart Editor" window and then click **Close** to return to the "Output – SPSS for Windows Viewer" window.

To remove the descriptive statistics (Std. Dev., Mean, and N) in the legend, follow these steps:

1. The chart must be in the "Chart Editor" window. If the chart is not in the "Chart Editor" window, double-click on the chart in the "Output – SPSS for Windows Viewer" window.
2. Click on any part of the Legend and it will be highlighted in blue.
3. Click delete and the Legend is removed.
4. If you are finished editing the chart, click **File** from the main menu bar within the "Chart Editor" window and then click **Close** to return to the "Output – SPSS for Windows Viewer" window.

To add numbers to the bars or change the color or the fill of the bars, follow the directions given above about editing bar graphs.

Stemplots

A stemplot (also called a stem-and-leaf plot) gives a quick picture of the shape of the distribution for a quantitative variable while including the actual numerical values in the graph. Stemplots work best for small numbers of observations that are all greater than zero.

We will work with the data shown in Table 1.2 in IPS for literacy rates in Islamic nations. The data can be retrieved from the Web site: www.whfreeman.com/ips5e. The SPSS for Windows Data Editor contains three variables called *country* (declared type string 12), *fperc*, and *mperc* (both declared type numeric 8.0).

To create a stemplot of this distribution, follow these steps:

1. Click **Analyze, Descriptive Statistics, Explore.** The "Explore" window in Figure 1.28 on the facing page appears.
1. Click *fperc,* then click ▸ to move *fperc* into the "Dependent List" box. Repeat for *mperc.*
2. By default, the "Display" box in the lower left corner has "Both" selected. Click **Plots** instead.
3. Next click **Plots** (located next to the "Options" button). The "Explore: Plots" window in Figure 1.29 appears.
4. Click **None** within the "Boxplots" box. Be sure that a ✓ appears in front of "Stem-and-leaf" within

the "Descriptive" box.

5. Click **Continue** and then click **OK.**

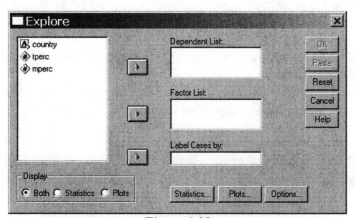

Figure 1.28

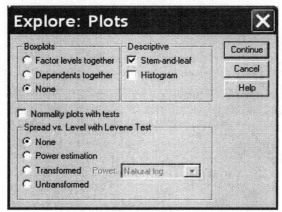

Figure 1.29

Part of the resulting SPSS for Windows output is shown in Table 1.1.

fperc Stem-and-Leaf Plot		*mperc* Stem-and-Leaf Plot	
Frequency	Stem & Leaf	Frequency	Stem & Leaf
1.00	2 . 9	1.00 Extremes	(=<50)
2.00	3 . 18	2.00	6 . 88
1.00	4 . 6	2.00	7 . 08
.00	5 .	4.00	8 . 3459
3.00	6 . 033	5.00	9 . 22456
4.00	7 . 0118	3.00	10 . 000
3.00	8 . 256		
3.00	9 . 999		

Stem width:	10	Stem width:	10
Each leaf:	1 case(s)	Each leaf:	1 case(s)

Table 1.1

More On Histograms

Example 1.7 and Table 1.3 present data on IQ test scores for 60 5th-grade students. We can explore these data for their "bell shape.

1. Open the file called *Table1.3.sav*. This data set includes one variable called *iq*.
2. Click **Graph, Histogram** as we did earlier.
3. In the "Histogram" box move the variable *iq* into the "Variable" box. Add titles if you wish. Then, click in the "Display normal curve" box. See Figure 1.30 shown below.
4. Click **OK**.

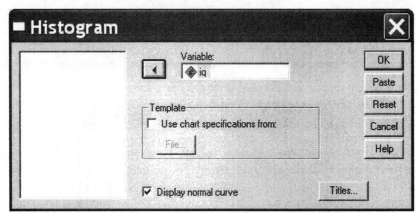

Figure 1.30

The SPSS for Windows output is shown in Figure 1.31.

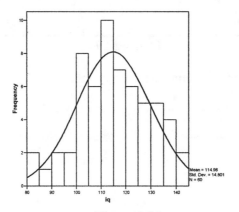

Figure 1.31

Examining Distributions

In any graph of data we look for the overall pattern and then describe shape, center, and spread as well as check for outliers. Look again at Figure 1.31. What is the overall shape? Where is the approximate center? How would you describe the spread?

You'll notice that the data shown above closely conform to the shape of a normal "bell" curve. The middle point on the histogram is at about 110. The lowest IQ in our data set is about 80 and the highest about 150, giving us an estimate of the spread.

Dealing With Outliers

Look at Example 1.9 in IPS. Here we have some data for breaking strength for the bond between wires and a semiconductor wafer. Using SPSS for Windows create a histogram such as the one in Figure 1.32. The data can be retrieved from the text Web site at www.whfreeman.com/ips5e. Can you explain the outliers? What would you do about them? What does the variability in breaking strength tell you about the manufacturing process? Is there need for greater quality control in the manufacturing process? Some of these questions are addressed in IPS. Brainstorm about answers to the questions that are not answered there. Can you think of additional questions to ask?

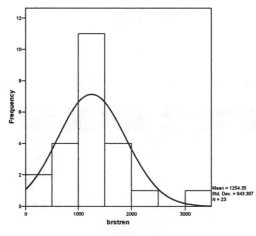

Figure 1.32

Time Plots

Many interesting data sets are time series, measurements of a variable taken at regular intervals over time. When data are collected over time, it is a good idea to plot the observations in time order. A time plot puts time on the horizontal scale of the plot and the quantitative variable of interest on the vertical scale.

Example 1.10 and Table 1.4 in IPS give the "Yearly discharge of the Mississippi River in cubic kilometers of water from 1954 to 2001". Make a time plot of these data. The data can be retrieved from the Web site: www.whfreeman.com/ips5e.

The SPSS for Windows Data Editor contains two variables called *year* and *discharg* (both declared numeric 8.0). Then create a histogram of discharge as shown in Figure 1.33 on the next page. Look for shape, center, spread, and outliers.

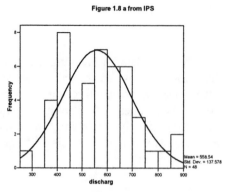

Figure 1.33

To create a time plot for this data set, follow these steps:

1. Click **Graphs** and then click **Sequence.** The "Sequence Charts" window shown in Figure 1.34 appears.

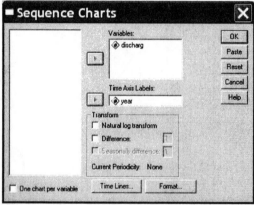

Figure 1.34

2. Highlight *discharg,* then click ▸ to move them into the "Variables" box.
3. Click *year* and then click ▸ to move *year* to the "Time Axis Labels" box.
4. Click **OK.**

Figure 1.35 on the next page is the resulting SPSS for Windows output. To edit the *x* and *y* axes, consult the directions in the Editing Histograms section. The time plot shows a gradual increase in discharge from the Mississippi River over the period from 1954 to 2001.

1.2 Describing Distributions With Numbers: Center And Spread

Table 1.10 in IPS gives us data on fuel economy in MPG for 2004 vehicles. Using numbers, we want to compare Two-seater cars to Mini-compact cars and city mileage to highway mileage. The following examples do not include graphical illustrations, however, it is good practice to do the visual summary to check for the "bell" shape using graphs before looking at the numerical summaries.

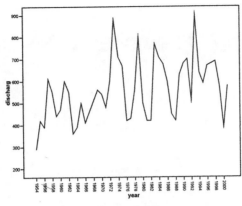

Figure 1.35

Measuring The Center

To get the Mean using SPSS for Windows (using the variable *twohwy* for this example), do the following:

1. Click **Analyze, Descriptive Statistics,** and **Descriptives** and the "Descriptives" window shown in Figure 1.36 will appear.

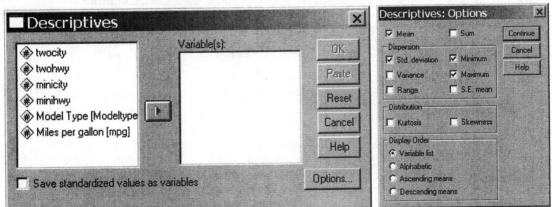

Figure 1.36   **Figure 1.37**

2. Click on *twohwy* and then click ▸ to move the variable into the "Variable(s)" box.
3. Click the "Options" tab and the "Descriptives: Options" window shown in Figure 1.37 will appear.
4. To only have the mean displayed, make sure there is a ✔ in the box beside "Mean" and uncheck any other boxes. Click **Continue** and then click **OK**. Your output will look like that in Table 1.2.

Descriptive Statistics

	twohwy	Valid N (listwise)
N	34	34
Mean	25.18	
Std. Deviation	8.730	

Table 1.2

To get the median value, together with other interesting statistics including a boxplot, follow these steps:

1. Click, **Analyze, Descriptive Statistics,** and **Explore** and the "Explore" window shown in Figure 1.38

will appear.

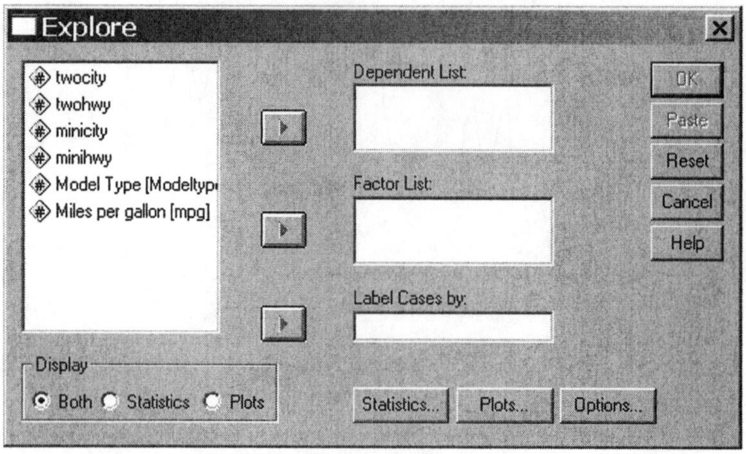

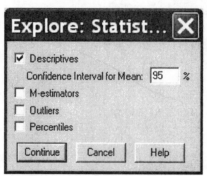

Figure 1.38

2. Click on *twohwy* and then click ▸ to move the variable into the "Dependent List" box.
3. Click on the "Statistics" tab and the "Explore: Statistics" window shown in Figure 1.39 will appear.
4. Make sure there is a ✔ in the boxes beside the statistics you desire. Click **Continue**.

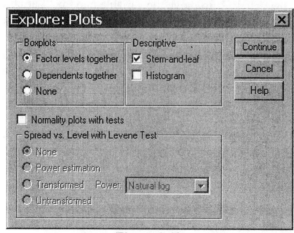

Figure 1.39

5. Click on the "Plots" tab and the "Explore: Plots" window shown in Figure 1.40 will appear.

Figure 1.40

6. Uncheck the ✔ in the box beside "Stem-and-leaf". Click **Continue** and then click **OK.**

Tables 1.3 to 1.5 are some of the statistics that will appear in the SPSS output.

Descriptives

			Statistic	Std. Error
twohwy	Mean		25.18	1.497
	95% Confidence Interval for Mean	Lower Bound	22.13	
		Upper Bound	28.22	
	5% Trimmed Mean		24.32	
	Median		24.50	
	Variance		76.210	
	Std. Deviation		8.730	
	Minimum		13	
	Maximum		66	
	Range		53	
	IQR		7	
	Skewness		3.070	.403
	Kurtosis		14.476	.788

Table 1.3

Percentiles

		Percentiles						
		5	10	25	50	75	90	95
Weighted Average(Definition 1)	twohwy	14.50	16.00	21.50	24.50	28.00	31.50	40.50
Tukey's Hinges	twohwy			22.00	24.50	28.00		

Table 1.4

Extreme Values

			Case Number	Value
twohwy	Highest	1	10	66
		2	21	32
		3	30	32
		4	31	31
		5	19	29(a)
	Lowest	1	12	13
		2	11	15
		3	8	16
		4	7	16
		5	14	17

a Only a partial list of cases with the value 29 are shown in the table of upper extremes.

Table 1.5

Measuring The Spread

A boxplot (see Figure 1.41) is a good way to assess center and spread visually. SPSS for Windows will also give you numerical values called the five-number summary together with your boxplot. These five numbers include the minimum and maximum values together with the median and first and third quartiles. Review the output shown above and locate the values for the five number summary for Two-seaters on the highway. Confirm them with the numbers shown in IPS. Now repeat the process for highway mileage for Mini-compacts. Did you get the same values as shown in the text?

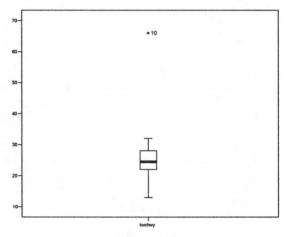

Figure 1.41

The center horizontal line in the boxplot shown above represents the median value. The upper and lower horizontal lines that outline the box represent the first quartile (lower line) and third quartile (upper line). If your data are normally distributed (take the form of a bell shape) then the median should be near the vertical middle of the box. The vertical lines extending above and below the box indicate the spread. Notice that car number 10 in our list (the Honda Insight) is a high outlier. If you scan the data in Table 1.10 you will see that highway mileage for this car is 66 MPG. The next highest is 28 MPG for several Two-seater models. The Honda Insight is clearly an outlier on the boxplot shown above. Why does the Honda Insight get such good mileage that it stands out from similar models in its class? Do a little research to answer this question.

How can we get a visual comparison of the mileage for both model types and for city and highway mileage? Follow these steps.

1. Click **Analyze, Descriptive Statistics,** and **Explore**. Move *mpg* to the "Dependent List" box and *modeltype* to the "Factor List" box. Your screen will now look like Figure 1.42 shown on the following page.
2. Click on the "Plots" tab and specify "Factor levels together" in the Boxplot box and uncheck "Stem-and-leaf" as shown in Figure 1.43.
3. Click **Continue** and then click **OK**.

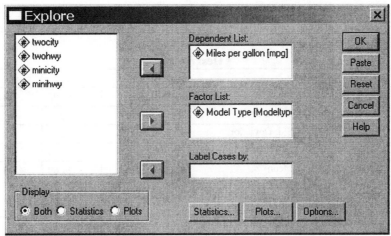

Figure 1.42

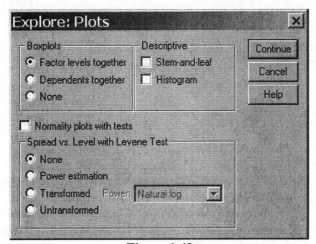

Figure 1.43

Compare your output (shown in Figure 1.44) to Figure 1.17 in IPS.

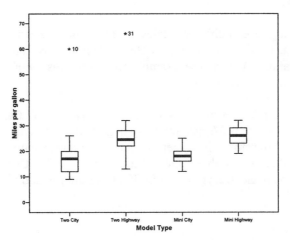

Figure 1.44

The most commonly used numerical measure of spread is the Standard Deviation. Look at Table 1.6.
This is an excerpt from the SPSS output that we are given when we ask for Descriptive Statistics.

Descriptives

		Statistic	Std. Error
twocity	Mean	18.08	3.665
	Variance	174.577	
	Std. Deviation	13.213	
	Interquartile Range	9	
twohwy	Mean	24.69	3.706
	Variance	178.564	
	Std. Deviation	13.363	
	Interquartile Range	11	
minicity	Mean	18.38	.971
	Variance	12.256	
	Std. Deviation	3.501	
	Interquartile Range	5	
minihwy	Mean	26.00	1.062
	Variance	14.667	
	Std. Deviation	3.830	
	Interquartile Range	6	

Table 1.6

Look for the values for the IQR, standard deviation, and variance. What is the value for the mean? Why
do we generally choose the mean and standard deviation to describe our distributions? Why do we use
graphical displays in conjunction with these two numerical measures?

For normally distributed data, the mean and standard deviation are the most commonly used statistics.
For skewed distributions, we generally use the five number summary. Plotting your data, something you
should always do, will help you to choose the appropriate numerical summary.

Linear Transformations

We can record variables using several different scales. For example, we can measure distance in miles or
kilometers. Using SPSS it is relatively easy to make the transformations from one unit of measurement to
another (a linear transformation). We can use Example 1.20 as an illustration. The data can be retrieved
from the Web site: www.whfreeman.com/ips5e.

To transform Mary's weights in ounces to grams, click **Transform**, **Compute**. In the Compute Variable
box, name your new variable (mine is *Marygr* for Mary in grams). Then click the variable *Mary* into the
Numeric Expression box. To transform these measurements to grams, we multiply each by 28.35 so add
the multiplication sign (*) after the variable name and then type in 28.35. Click **OK**. The transformed
data will appear as the new variable *Marygr* in the SPSS Data Editor window. (See Figure 1.45 on the
facing page.)

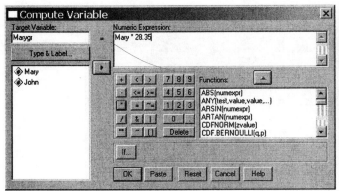

Figure 1.45

To confirm that the statistics don't change (except that they are now 28.35 times higher), do descriptive statistics on **Mary** and **Marygr**. Excerpts taken from the SPSS output are shown in Table 1.7. Then compare Mary's measurements in grams to John's. First, however, let's check our data for normality by doing a histogram for each with a normal curve overlaid. Keep in mind that these are very small data sets. The outputs for John and Mary in grams are shown in Figure 1.46. Notice that in the lower right corner of each histogram you can find the mean, standard deviation, and sample size.

Descriptives

		Statistic	Std. Error
Mary	Mean	1.1200	.03701
	Variance	.007	
	Std. Deviation	.08276	
	Interquartile Range	.15	
Marygr	Mean	31.7520	1.04933
	Variance	5.505	
	Std. Deviation	2.34638	
	Interquartile Range	4.39	

Table 1.7

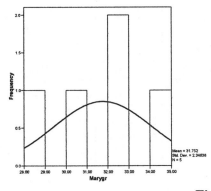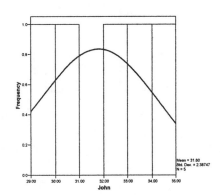

Figure 1.46

If you stack the data as you did previously in the example for the car types, we can do side-by-side boxplots. To stack the data and make the boxplots, follow these steps.

1. Start by entering two new variables in your data set called **Individual** and **Grams** as shown in

Figure 1.47.

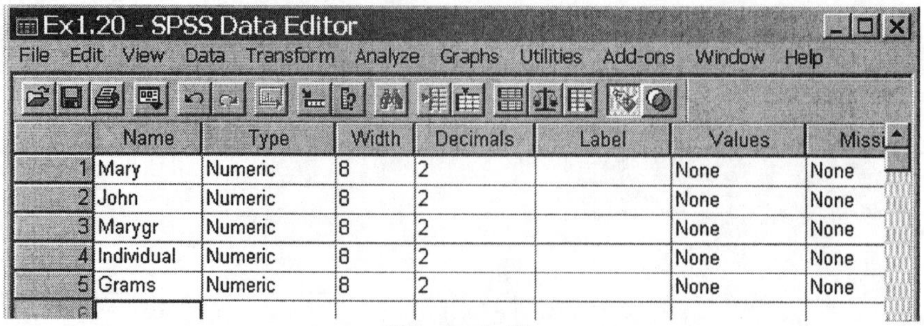

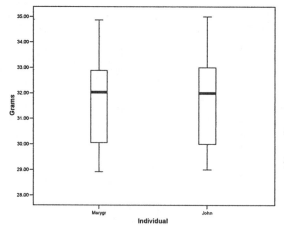

Figure 1.47

2. For *Individual*, change the decimal places to 0 and add value labels (refer to Chapter 0 for how to add labels). In this example, we used 1 to represent *Marygr* and 2 to represent *John*. Now, stack all of your data into one column by highlighting the cells you want to copy, copying them, then pasting them into the new column called *Grams*. Beside each of the weight entries in the *Grams* column, place a 1 in the *Individual* column. See Figure 1.48 for Mary's data in grams. John's data are below Mary's.

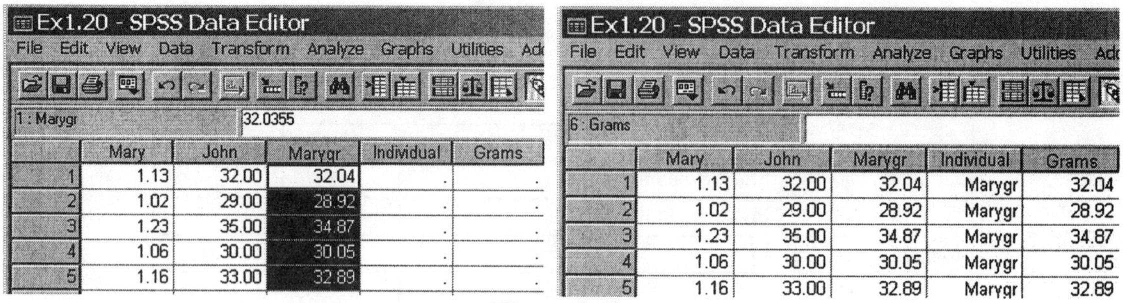

Figure 1.48

3. Do the same for John's data but change the number in the *Individual* column to 2. Now create the side-by-side boxplots.

Figure 1.49

From Figure 1.49 above, once we have converted Mary's measurements to grams, is there any reason to think that there is any difference in their measurements of the weights of newborn pythons?

1.3 Density Curves And Normal Distributions

When we made a histogram earlier with a normal curve laid over it, the normal curve represents a density curve. It is a smooth approximation of the bars of a histogram with a total area under the curve equal to 1 or 100% of the area. See, for example, Figure 1.21a in IPS for a large sample of grade equivalent vocabulary scores. The area under the curve that is shaded in Figure 1.23a in IPS is approximately 30% of the total area under the curve. Thus the proportion or percent of students having a vocabulary score of less than 60 is 30%. This is the same as saying that the proportion of students with a vocabulary score of less than 6 is 0.3.

The normal curve is a specific type of density curve. In a normal curve, the mean and median scores are the same and each half is a mirror image of the other (it is symmetric). The characteristics of a density curve are given by its mean (center) and standard deviation (spread). You can thus locate the middle by drawing a vertical line that bisects the two halves. You can also draw in the first standard deviation by drawing in a vertical line at the point where the curve changes from convex to concave.

Cumulative Density Functions

We can divide every normal curve into about 6 parts based on the standard deviation. We can use Exercise 1.30 to illustrate the most important characteristic of all normal curves. The data can be retrieved from the Web site: www.whfreeman.com/ips5e. This characteristic is described as the 68-95-99.7 rule. Look at Figure 1.50 below that is a histogram of the ph levels for 105 samples of rainwater. The mean acidity (ph level) is 5.4 and the standard deviation is .54. Consider these numerical descriptions of the distribution of acidity in rainwater and be sure that you understand the statement that 95% of rainwater samples will have a ph level between 4.32 and 6.48. We can use our SPSS histogram or descriptive statistics to give us this information. Recall that SPSS prints out the mean and standard deviation beside each histogram when you have also asked it to overlay the normal curve. To have 95% of the area under the curve, we start at the center (mean = 5.4) and first subtract 2 standard deviations (.54*2) to get 4.32. Then we add the same amount to the mean (5.4 + 2*.54) to get 6.48. Thus we can make the statement that 95% of rainwater samples will have a ph level between 4.32 and 6.48.

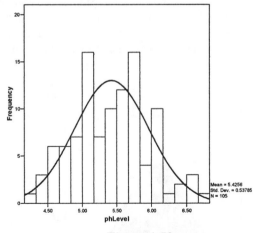

Figure 1.50

But, what if we wanted the percentage of ph levels for portions of the area under the curve that are not even multiples of the standard deviation? Or are not located in the middle in a symmetric way? Table A in your text gives you close approximations for the most commonly used areas under the curve. SPSS will calculate these for you also.

The first step in the process is to standardize the values, that is, to translate them into their equivalents with a mean of 0 and a standard deviation equal to 1. The data is already entered in your spreadsheet under the values that SPSS automatically labeled *zphLevel.* For future reference on how to do this in SPSS using the Analyze, Descriptives menu options, follow the steps below.

1. Click **Analyze, Descriptive Statistics,** and **Descriptives.** The "Descriptives" window appears (see Figure 1.37).
2. Move *phLevel* into the "Variable(s):" box and click on the box beside "Save standardized values as variables". Click OK. Your spreadsheet will now have an added column with z-scores in it as shown in Figure 1.51.

Exercise 1.30 - SPSS Data Editor

File   Edit   View   Data   Transform   Analyze   Graphs   Utilities   Add-ons   Window   Help

2:                   Redo

	phLevel	ZphLevel	var	var	var	var	var
2	4.75	-1.25614					
3	4.94	-.90288					
4	5.10	-.60540					
5	5.19	-.43807					
6	5.41	-.02904					
7	5.55	.23125					
8	5.68	.47296					
9	5.81	.71466					
10	6.03	1.12369					
11	6.43	1.86739					
12	4.38	-1.94406					
13	4.76	-1.23755					
14	4.96	-.86570					

Data View   Variable View

Redo   SPSS Processor is ready

Figure 1.51

Now, let's ask SPSS to assist us in solving a problem similar to Example 1.29 in IPS but using our variable *phLevel.* Look again at the histogram for *phLevel.* We can ask what proportion (or percent) of the ph levels fall between 4.5 and 6.0. Recall from the discussion in IPS that density curves are cumulative frequency curves, that is they "add up" from left to right. The piece that we want is shaded in Figure 1.52. This means that we cumulate or "add up" to 6.0, then "add up" to 4.5. To get the cumulative density function for the shaded area under the curve, we then subtract the small piece on the left from the larger piece. The steps are outlined on the next page.

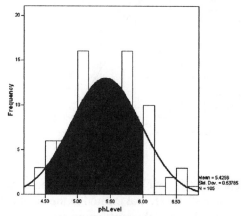

Figure 1.52

1. Check your data set to be sure there are actual values of 4.5 and 6.0. In this case there are. Recall that we calculated the z-values earlier and saved them as the variable *zphLevel*. Click **Transform** and then **Compute** as shown in Figure 1.53.

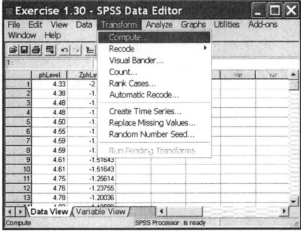

Figure 1.53

2. The "Compute Variable" window will appear as shown in Figure 1.54. Fill in a variable name for the CDF values that we are about to compute, for example *CDF*. Then click the scroll bar in the "Functions:" box and scroll down until you see **"CDFNORM[?]"**. Click to highlight, then click the up arrow in the box beside the "Functions:" box and you will now see **"CDFNORM[?]"** in the "Numeric Expression" box. Click to highlight the question mark, and then click on the variable *zphLevel* and the box containing the right arrow and now you will see *zphLevel* in place of the question mark. Click **OK.**

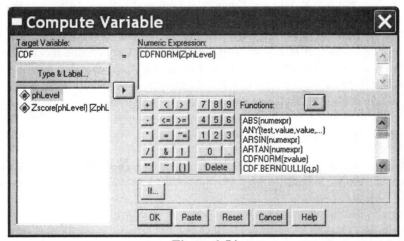

Figure 1.54

3. The SPSS output is shown in Figure 1.55 on the next page. Scan the column for the variables *phLevel* and *CDF*. Opposite the *phLevel* of 4.5 we find a *CDF* score of .04. Next to a *phLevel* of 6.0 we find a *CDF* score of .86.

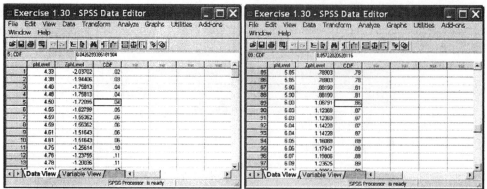

Figure 1.55

Recall that we were asked for the area under the curve between 4.5 and 6.0. Therefore to complete our calculations we subtract .04 from .86 to get a proportion equal to .82. Thus 82% of our rainwater samples have ph levels between 4.5 and 6.0.

Inverse Normal Calculations

Instead of calculating the proportion of observations in relation to specific scores, we may want to find specific scores that correspond to certain percentages. You can use Table A in IPS to do this or use built-in functions in SPSS. We will continue with ph levels in our rainwater samples to illustrate. Suppose, for example, that we want to know how high the ph level must be to place the sample in the top 10% of our observed ph levels.

1. First use the Analyze portion of the program to calculate the mean and standard deviation for the sample (or read it off of the histogram that you have already produced). For this example, the mean ph level for all of the samples is 5.43 and the standard deviation is .54. SPSS will prompt you for these values as you run the program.
2. Next, click **Transform, Compute**. Then give the "target Variable" a name such as *invcdf*. Now, search the "Functions:" box for **IDFNormal[p,mean,stdev].** Click on this function and then click the up arrow to move the function into the "Numeric Expression" box. Replace the p with .9 (for the top 10%), mean with 5.43, and stdev with .54. Click **Continue**.

The value that SPSS returns to you is 6.12. This means that to be in the top 10% of ph readings, a sample must have a ph greater than 6.12.

Explore the "Sort Cases" option under the "Data" menu to sort your samples from lowest to highest. In this set of rainwater samples, how many are greater than 6.12? By my count there are 8. Check this for yourself.

Normal Quantile Plots

Recall that we used graphical and numerical summaries of data to check for shape, center, and spread. While the visual inspection works well for us in some cases, there are more accurate ways to determine if the sample forms a normal distribution. One of the most useful tools for an accurate assessment of normality of shape is to produce a normal quantile plot. Follow these steps to produce the plot. Once again we will use our ph levels for the example.

1. Click **Graphs, Scatterplot**, "Simple" and then click **Define.** Move the variable *phlevel* into the box

for the "Y axis" and then move the variable *zphLevel* into the box for the "X axis". SPSS will provide the graph shown below in Figure 1.56.

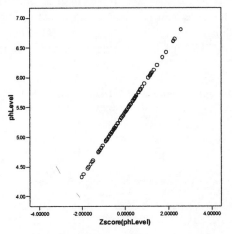

Figure 1.56

In Figure 1.56 we have plotted the observed scores for *phLevel* against their corresponding standardized (z) scores. When the points on such a plot fall on or close to a straight line, as ours do, we have a good indicator that the scores are normally distributed. Scan such a plot for outliers, especially in the high or low directions as described in IPS.

Exercises For Chapter 1

1. Most experts in the field of sleep and dreaming credit Eugene Aserinsky with the discovery of REM sleep. Shortly after this, William Dement published data showing some basic facts about dreaming. In 1960 he published his findings for the first eight subjects (Dement, W. (1960). The effect of dream deprivation. *Science, 131*, 1705-1707). Here are data adapted from that study:

Subject	1. Baseline Dream Time As A Percent Of Sleep Time	2. Number Of Dream Deprivation Nights	3. Number Of Awakenings		4. Percent Dream Time Recovery
			3a. First Night	3b. Last Night	
1	19.5	5	8	14	34.0
2	18.8	7	7	24	34.2
3	19.5	5	11	30	17.8
4	18.6	5	7	23	26.3
5	19.3	5	10	20	29.5
6	20.8	4	13	20	29.0
7	17.9	4	22	30	19.8
8	20.8	3	9	13	*

* Subject dropped out of study before recovery nights.

    a. Enter these data in an SPSS spread sheet. Use the * to represent the missing value in column 4.
    b. Draw a pie chart representing the baseline percentage of dream time for each subject.
    c. Draw a bar graph for the percent of dream time recovery and describe the overall pattern of the distribution. Are there any outliers?
    d. Draw a boxplot for the average percent of dream time. What is the mean and standard deviation (use the five number summary) for this variable? Describe what this means in terms of differences between subjects before the experimental data was collected.
    e. Draw a histogram for the percent dream time recovery and place a normal curve over it. What does this tell you about the percentage of time we spend in dream time recovery after dream deprivation?
    f. Make normal quantile plot for percent of dream time recovery. Does this give you the same answer as in part e?

2. Here is an example of outcomes for 25 tosses of a die:
    5 4 6 4 5 1 1 3 6 1 3 4 4 3 5 1 4 3 2 3 4 4 5 6 2
    a. Make a bar graph for these outcomes.
    b. Do these data make a normal curve? Would you expect them to?
    c. For this set of outcomes, what was the most common number to come up? Would it be the same if you repeated this experiment? Why or why not?

3. Here are some fictional starting salaries (in thousands of dollars) for 40 new teachers:
    34.2 28.9 37.3 26.6 17.4 31.0 22.0 24.4 29.9 25.8 34.3 27.1 25.5 25.3 22.6 29.8 29.7 26.3 23.8 35.2 28.4 34.8 21.7 26.5 21.5 26.1 33.9 30.5 29.0 32.1 36.8 32.8 23.0 19.9 30.3 24.7 32.6 22.6 28.0 24.0
    a. Do descriptive statistics for these data. Double-check to make sure that you have entered all the data correctly and that you have 40 entries.
    b. Make a histogram of these data with a normal curve lying over it. Use intervals of 4 ($4000). What does this tell you about starting salaries for our fictional teachers? Describe the overall shape. Are there any outliers?

  c. What are the mean and median starting salaries?

  d. What is the standard deviation for these fictional salaries? How do you interpret that number?

  e. If each of these teachers receives a cost of living increase next year of 3.2%, what will their salaries be at that time?

  f. Repeat steps a through d for these new salary figures. What does this tell you?

4. During a 3-month period, a real estate firm made 27 single-family residential housing sales. The selling prices of these houses were: 72800 84000 69500 102000 84500 98800 61500 58000 89500 64500 118900 89000 128500 78000 94500 112000 89500 56700 95000 67800 101500 91300 88800 51800 122000 57500 127300

  a. Make a histogram with 6 classes of price with a normal curve overlaid.

  b. Make a histogram with 8 classes of price with a normal curve overlaid.

  c. Create the five number summary for these data.

  d. Make a boxplot for the full data set. Note where each number from the five number summary is placed in the boxplot.

  e. Describe the distribution of selling prices in terms of shape, center, and spread. Are there any outliers?

  f. What is the IQR for these data?

  g. Create a normal quantile plot for these data. Does it confirm your interpretation of the shape of this distribution?

5. Imagine that you have administered a math anxiety test to 100 students taking their first statistics course. Here are the (fictional) scores:

   51 50 50 51 50 50 48 49 46 50 46 45 46 47 46 36 36 38 37 36 57 54 59 57 58 49 48 49 48 45 48 46 48 46 48 34 35 44 64 49 49 47 43 47 46 47 48 43 48 46 46 48 46 48 37 37 37 37 59 59 46 47 47 45 44 45 48 48 48 47 47 49 47 35 48 49 47 45 47 44 48 57 47 46 47 46 47 46 45 47 45 47 75 48 48 46 48 45 46 47

  a. Make a stem-and-leaf of these scores.

  b. Create the five number summary. What are the highest and the lowest scores?

  c. Describe the shape, center, and spread of this distribution of anxiety scores.

  d. Make a histogram with an normal curve overlaid. Does this graphical representation of the anxiety scores confirm your analysis of shape, center, and spread?

  e. What is the mean and standard deviation for the anxiety scores? What measure did you pick to assess the center, and spread, and why?

6. The mean, median, standard deviation, and normal probability curve are all related to each other. Enter the following data in an SPSS spreadsheet and then answer the questions that follow.

   110 105 105 100 100 100 95 95 90

  a. Draw a histogram of the data with a normal curve overlaid. Is the curve unimodal (only one peak) and symmetrical?

  b. Calculate the mean and median. How do these scores compare to one another?

  c. What percentage of individual scores are greater than 100? How do you know this?

  d. What is the standard deviation for this set of scores? What percentage of scores falls between the mean and plus or minus one standard deviation?

  e. Sketch in the standard deviations by eye on your histogram with the normal curve overlying it.

7. Open your data set for the math anxiety scores and calculate the following using the CDF function:

  a. What score represents the position where 50% of the anxiety scores are greater than this score and 50% are less than this score?

  b. What percentage of the anxiety scores fall between 40 and 50?

  c. What percentage of the anxiety scores fall between 40 and 60?

    d.   What percentage of the anxiety scores fall between 35 and 75?

    e.   Calculate the standard deviation. Prove to yourself that approximately 15% of the anxiety scores are greater than this value.

    f.   What percent of the scores fall between -2s and +2s?

8.   The following data are representative of the amount of snow in inches that fell in upper Minnesota each year for the past 50 years:

    18 22 25 38 45 46 48 55 56 57 58 62 63 65 66 67 68 72 73 74 76 78 79 80 80 81 82 84 86 86 88 90 91 91 91 92 93 97 98 100 102 103 105 108 110 115 118 119 125 138

    a.   Make a boxplot and five number summary for these data.

    b.   Describe the shape center, and spread from the boxplot.

    c.   Make a histogram with a normal curve overlaid.

    d.   What is the mean and standard deviation for these data?

    e.   In what percentage of the last 50 years has snowfall been 100 inches or more?

    f.   What percentage of years saw less than 60 inches of snow?

9.   Children often use the phrase "and then" when recalling stories. Here are some representative data for the number of "and then's":

    10 11 12 15 15 15 16 16 16 16 17 17 17 17 17 17 18 18 18 18 18 18 18 18 18 18 19 19 19 19 19 19 19 20 20 20 20 20 20 20 21 21 21 22 22 23 23 24 31 40

    a.   Make a histogram for these data and place a normal curve over it.

    b.   Describe the shape center, and spread for these data.

    c.   Make a stem-and-leaf for these data. What problem do you observe?

    d.   What percentage of children use less than 16 "and then's"?

    e.   What score corresponds to the point at which 75% of the number of "and then's" fall below this score?

10. Adults use considerably fewer "and then's" than children do. Here are some representative scores:

    1 3 4 5 5 7 7 7 7 8 8 8 8 9 9 9 9 9 9 9 9 10 10 10 10 10 10 10 10 11 11 11 11 11 11 12 12 12 12 12 13 14 14 14 14 15 15 15 16 16 17

    a.   Make a histogram for these data and place a normal curve over it.

    b.   Describe the shape center, and spread for these data.

    c.   What score corresponds to the point at which 75% of the number of "and then's" for adults fall below this point?

    d.   What score corresponds to the point at which 40% of the number of "and then's" fall below this score?

    e.   Make side by side boxplots for the adult and the children data. What do you observe about maximum and minimum scores for each group? What about the mean score for each group?

Chapter 2. Looking at Data — Relationships

Topics covered in this chapter:

This chapter introduces analysis of two variables that may have a linear relationship. When two variables are measured for the same individual (e.g. height and weight) and the values of one variable tend to occur with a set of values for the other variable, they are said to be **associated**. In analyzing two quantitative variables, it is useful to display the data in a **scatterplot** and determine the **correlation** between the data. A scatterplot is a graph that puts one variable on the x axis and the other on the y axis, and it is used to determine whether an overall pattern exists between the variables. Traditionally the **explanatory** variable is placed on the x axis and the **response** variable on the y axis. The correlation measures the strength and direction of the linear relationship, and the least-squares regression line is the equation of the line that best represents the data.

2.1 Scatterplots

Humans, in general, are interested in the patterns of nature. Scatterplots present a visual display of the way variables "go together" in the world. In IPS Example 2.5, scientists take a look at the relationship between the size of an animal population and the number of predators there are in an area. The data can be retrieved from the Web site: www.whfreeman.com/ips5e.

To generate a scatterplot, follow these steps:

1. Click **Graphs, Scatter, Simple**, and then **Define.** The "Simple Scatterplot" window in Figure 2.1 appears.
2. Click *perch,* then click ▸ to move *perch* into the "X Axis" box.
3. Click *proportion,* and then click ▸ to move *proportion* into the "Y Axis" box.
4. Click **OK.**
5. Using the editing instructions presented before, edit the x and y axis labels to match those in Figure 2.4 from IPS.

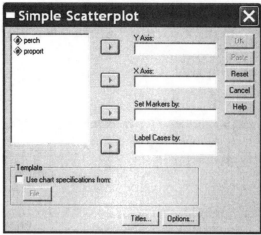

Figure 2.1

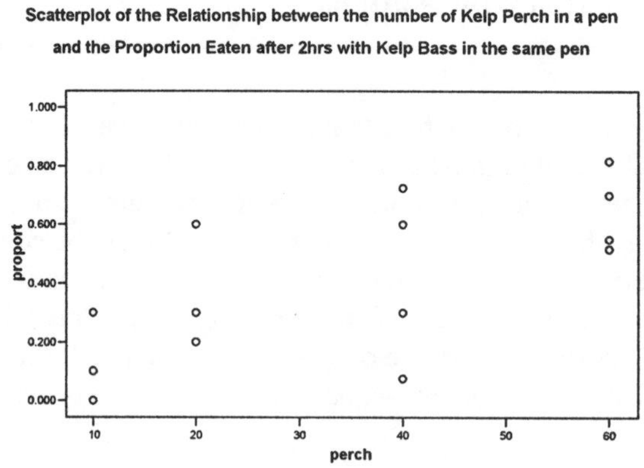

Scatterplot of the Relationship between the number of Kelp Perch in a pen
and the Proportion Eaten after 2hrs with Kelp Bass in the same pen

Figure 2.2

Figure 2.2 is a scatterplot of these data. ***Perch*** is considered the explanatory variable and ***proportion*** the response variable. We can add the mean for each fish pen to get a clearer picture of the relationship between these variables.

To do so, follow these steps:

1. First get the mean of each variable following the steps outlined in Chapter 1.
2. Then enter the data in the last row of each column as shown in Figure 2.3. As you can see in row 17 of the SPSS Data Editor window, the means for each have been added (***perch*** = 33; ***proportion*** = .399).
3. Make a scatterplot of the data by following the instructions in the scatterplot section above.
4. To make the point representing the means stand out, you can change the color or pattern of it by editing the chart.
5. To edit the chart, it must be in the "Chart Editor" window. If the chart is not in the "Chart Editor" window, double-click on the chart in the "Output – SPSS for Windows Viewer" window.
6. Click on any point in the chart. You'll notice that all of the points will be highlighted in blue.
7. Click on the point you want to change to have only it highlighted in blue.

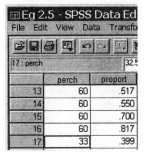

Figure 2.3

8.   Double-click the point to bring up the "Properties" window shown in Figure 2.4.

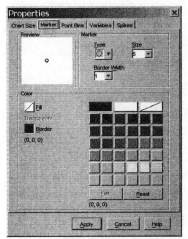

Figure 2.4

9.   Here you can change the type, size, and color of the marker. Click on "Type" and you can select from a variety of shapes as shown in Figure 2.5.

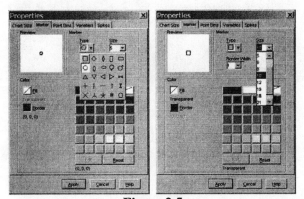

Figure 2.5

10. For this example, we used a square and changed the size of the marker to 10 by selecting 10 from the drop down menu in the "Size" box. See Figure 2.5

11. To change the color of the marker, click "Fill" and then select a color. For this example, we used black.
12. Click **Apply** and then click **Close.**

Figure 2.6 is the resulting SPSS for Windows Output.

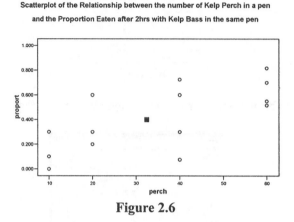

Figure 2.6

Adding Categorical Variables To Scatterplots

Table 2.3 in IPS gives us world record times for the 10,000 meter run for men and women. Enter these data in a spreadsheet showing year, time (sec), and gender. Then make a scatterplot of these data showing times for men and times for women as different markers. The steps are shown below.

1. Click **Graphs,** click **Scatter,** click **Simple** and then click **Define.** The "Simple Scatterplot" window in Figure 2.7 appears.

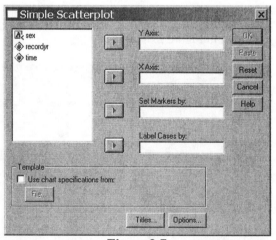

Figure 2.7

2. Click *recordyr,* and then click ▸ to move *recordyr* into the "X Axis" box.
3. Click *time,* and then click ▸ to move *time* into the "Y Axis" box.
4. Click *sex,* and then click ▸ to move *sex* into the "Set Markers By" box. Add titles if you wish. Click **OK.**

The scatterplot visualizing the relationship between the year of the race and the time in seconds for Males and Females is shown in Figure 2.8. The data for Males is shown by open circles and for Females by filled circles. As expected, males are generally faster than females. Thus, the data for Males tends to be in the lower left corner of the plot while the data for Females tends to be in the upper right quadrant. There appears to be a strong negative relationship between the record year and the time (seconds) for both males and females with the males having faster times than the females overall. Notice that the decrease in time is greater in females than in males. What does this tell you?

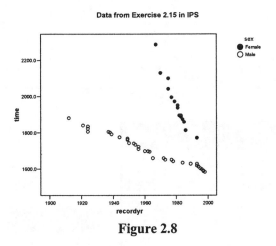

Figure 2.8

2.2 Correlation

While the scatterplot gives us visual information about the relationship between two quantitative variables, correlation gives us the numerical value for the relationship. Using the data from Exercise 2.14, make a scatterplot of the relationship between speed and fuel consumption for the British Ford Escort. Describe the nature of the association between these variables from the visual output. Then compute the correlation (numerical description of the relationship) using the directions shown below.

To obtain the correlation follow these steps:

1. Click **Analyze, Correlate,** and then **Bivariate,** as illustrated in Figure 2.9 on the following page.
2. Click *speed,* then click ▸ to move *speed* into the "Variables" box.
3. Click *fuel,* and then click ▸ to move *fuel* into the "Variables" box also. See Figure 2.10.
4. If you are interested in obtaining descriptive statistics for the variables, click the **Options** box, click **means and standard deviations,** and then click **Continue.** See Figure 2.11.
5. Click **OK.**

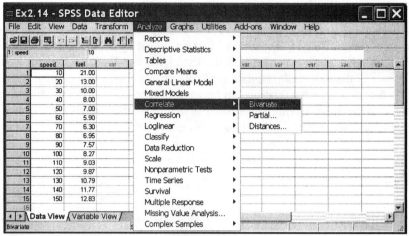

Figure 2.9

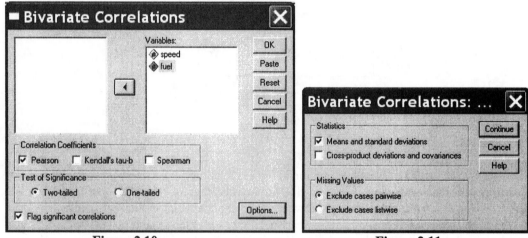

Figure 2.10   **Figure 2.11**

The correlation and the descriptive statistics for speed and fuel are shown in Table 2.1 on the facing page. The descriptive statistics show the means and standard deviations for the data. The correlation is negative at −.779 and it is significant, $p = .000$.

2.3 Least-Squares Regression

The example used in this section is taken from IPS Example 2.9 that addresses the question "Does fidgeting keep you slim?" The data can be retrieved from the Web site: www.whfreeman.com/ips5e. The scatterplot for these data is shown in Figure 2.12 using nonexercise activity increase (***neaincr***) as the explanatory (*x*) variable and fat gain in kilograms (***fatgain***) as the response (*y*) variable. Notice that there is a moderately strong, negative, linear correlation between these two variables. This scatterplot was produced using the instructions shown above. Notice that there is a negative correlation between activity and gain: the more active you are, the less you gain.

Descriptive Statistics

	Mean	Std. Deviation	N
fatgain	2.388	1.1389	16
neaincr	324.75	257.657	16

Correlations

		fatgain	neaincr
Pearson Correlation	fatgain	1.000	−.779
	neaincr	−.779	1.000
Sig. (1-tailed)	fatgain	.	.000
	neaincr	.000	.
N	fatgain	16	16
	Neaincr	16	16

Table 2.1

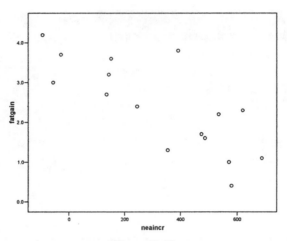

Figure 2.12

To obtain statistics such as the correlation coefficient, y intercept, and slope of the regression line, follow these steps:

1. Click **Analyze, Regression, Linear.** The "Linear Regression" window in Figure 2.13 appears.
2. Click *neaincr,* then click ▸ to move *neaincr* into the "Independent(s)" box.
3. Click *fatgain,* and then click ▸ to move *fatgain* into the "Dependent" box.
4. If you are interested in obtaining descriptive statistics for the variables, click **Statistics.** In the "Linear Regression Statistics" window shown in Figure 2.14, click **Descriptives, Continue** and **OK.**

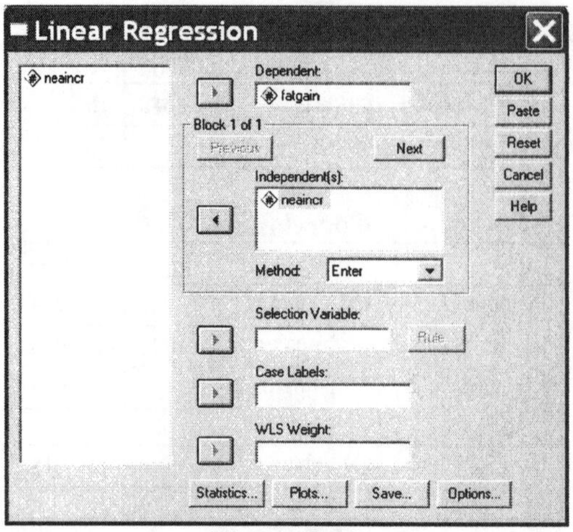

Figure 2.13

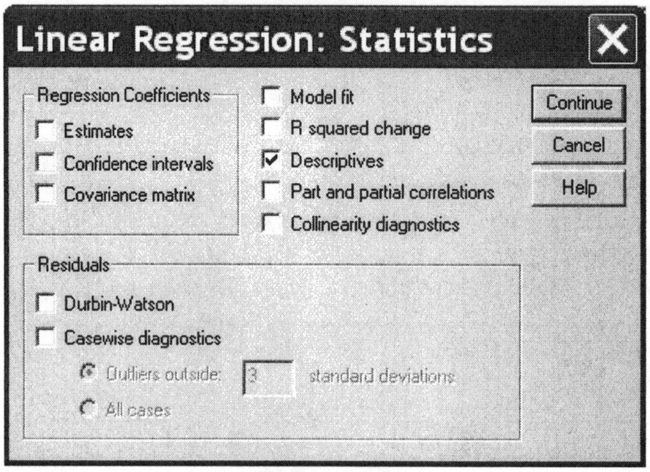

Figure 2.14

Table 2.2 is part of the SPSS for Windows output that show the (**R**), r^2 (**R Square**), the y-intercept (labeled **Constant**), and the slope (labeled **activity**).

Model Summary

Model	R	R Square	Adjusted R Square	Std. Error of the Estimate
1	.779(a)	.606	.578	.7399

a Predictors: (Constant), neaincr

Coefficients(a)

Model		Unstandardized Coefficients		Standardized Coefficients		
		B	Std. Error	Beta	t	Sig.
1	(Constant)	3.505	.304		11.545	.000
	neaincr	-.003	.001	-.779	-4.642	.000

a Dependent Variable: fatgain

Table 2.2

From these tables, we can construct the regression equation as:

$$\text{Predicted } \textbf{\textit{fatgain}} = 3.505 + (-.003)(\textbf{\textit{neaincr}})$$

These coefficients (values of the intercept (**Constant**) and slope (**neaincr**)) are read from the "Unstandardized Coefficients" box in Table 2.3 and inserted in the appropriate places in the equation for a straight line.

Fitted Line Plots

To plot the least-squares regression line on the scatterplot, follow these steps:

1. Generate the scatterplot as shown in Figure 2.12.
2. When the scatterplot appears in the Output window, double-click inside the scatterplot to gain access to the Chart Editor.
3. Once in the Chart Editor, click on one of the data markers to highlight all of the data points. If you wish, you may change the properties of the markers and other features of the scatterplot by double-clicking a point and following the instructions mentioned above under the generating a scatterplot section. Otherwise click **Close**.
4. Click **Chart, Add Chart Element, Fit Line at Total**. See Figure 2.15.

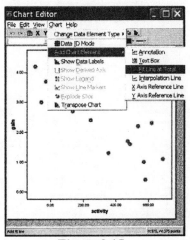

Figure 2.15

5. A "Properties" box will open as shown in Figure 2.16 on the following page. Be sure that **Linear** is clicked and then click **Close**.
6. Now click **Close** on the **Chart Window** and the fitted line will be added to the scatterplot as shown in Figure 2.17.

This fitted line is also called the **least-squares regression line** and it is used when making predictions based on the information in the scatterplot. Notice also that in the **fitted line plot** shown to the left in Figure 2.17 above, not all of the points are right on the line. That means there is scatter or variability in the relationship between these two variables.

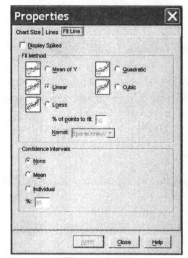

Figure 2.16

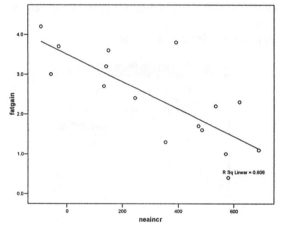

Figure 2.17

Predictions

We can use the regression line to **predict** the response *y* for a specific value of the explanatory *x* variable. In Example 2.10, we want to predict the fat gain for an individual whose nonexercise activity increases (***neaincr***) by 400 calories. We can make this prediction "by eye" from our fitted line plot. Locate the point on the *x* axis that represents 400 calories. Draw a vertical line up from that point to the fitted line. Now draw a horizontal line from that point to the y axis. The line ends just above 2 kg. See Figure 2.12 in IPS for an example.

It is more accurate to use the equation for our fitted line. Recall that we developed the equation as:

$$\text{Predicted } \boldsymbol{fatgain} = 3.505 + (-.003)(\boldsymbol{neaincr})$$

We can calculate the gain numerically by inserting 400 in place of the word ***neaincr*** in our formula. We can complete the calculations to get 2.13 kilograms of weight gain. Notice that this confirms our "by eye" estimate.

Avoid **extrapolation**, which is making predictions far outside the range of values shown on our scatterplot. The simplest way to avoid extrapolation is to do the "by eye" estimate first. Don't make estimates that are beyond the range of values shown on the *x* axis.

Correlation and Regression

Correlation and regression are closely related. When assessing correlation, it is not necessary to distinguish between the explanatory and response variables. However, this is critically important in regression since different regression lines (and hence predictions) are produced if you reverse the response and explanatory variables.

We will continue with Example 2.9 in this section. To begin, review Table 2.1 shown earlier in this chapter. You will notice that the correlation between *neaincr* and *fatgain* is the same as the correlation between *fatgain* and *neaincr* (−.779). When calculating correlation it does not matter which variable is considered the response or the explanatory variable. This is not the case for regression because the least-squares regression line and predictions made from it are dependent on the definitions of the variables.

In Figure 2.18 below, there are two graphs, one showing the regression line for predicting *fatgain* from *neaincr* and the other predicting *neaincr* from *fatgain*. These were created using the instructions in an earlier section for fitted line plots. Notice that the regression line is in a different place for each of the two fitted line plots. When predicting *fatgain* from *neaincr*, *fatgain* MUST be on the *y* axis. When predicting *neaincr* from *fatgain,* then *neaincr* MUST be on the *y* axis.

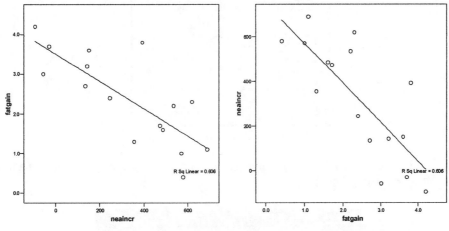

Figure 2.18

The value of the correlation, when squared, gives us a measure of the strength of the strength of a linear relationship. We saw that SPSS computed the correlation to be −.779 (see Table 2.2 earlier in this chapter). Notice also that we have the value for *r*-square in Table 2.2 right next to the value of *r*. The value for *r*-square is .606. This tells us that about 60% of the variability in *fatgain* is explained by its relationship with *neaincr*.

2.4 Cautions About Correlation And Regression

Residuals

A **residual** (or the **error** of prediction) is the difference between an observed value of the response variable and the value predicted by the regression line, written $residual = y - \hat{y}$. The primary purpose for analyzing residuals is to determine whether or not the linear model best represents a data set.

To generate the residuals, follow these steps:

1. Click **Analyze, Regression, Linear.**
2. Click *neaincr,* then click ▸ to move *neaincr* into the "Independent(s)" box.
3. Click *fatgain,* and then click ▸ to move *fatgain* into the "Dependent" box.
4. Click **Save,** then, in the "Residuals" box, click **Unstandardized** (see Figure 2.19).
5. Click **Continue,** and then **OK.**

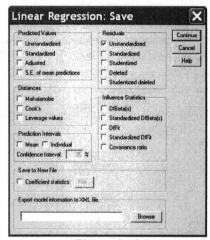

Figure 2.19

The residuals for this data set have been generated, saved, and added to the Data Editor in a new column labeled **"RES_1"**, as shown in Figure 2.20.

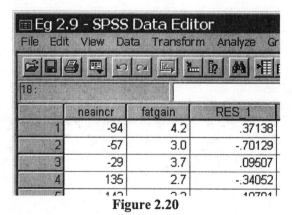

Figure 2.20

Residual Plots

Residual plots are a type of scatterplot in which the independent variable is generally on the x axis and the residuals are on the y axis. Such a plot helps us to assess the fit of a regression line. It is desirable for no pattern to exist on the residual plot, that is, for the plot to be an unstructured band of points centered around $y = 0$. If this is the case, then a linear fit is appropriate. If a pattern does exist on the residual plot, it could indicate that the relationship between y and x is nonlinear or that perhaps the variation of y is not constant as x increases. Residual plots are also useful in identifying outliers and influential observations.

To plot the residuals against the independent variable (*neaincr*) and to plot the reference line at $y = 0$ on this plot, follow these steps:

1. Click **Graphs, Scatter, Simple, Define.**
2. Click *neaincr,* and then click ▸ to move *neaincr* into the "X Axis" box.
3. Click *Unstandardized Residuals (RES_1)*, then click ▸ to move *Unstandardized Residuals (RES_1)* into the "Y Axis" box (see Figure 2.21).

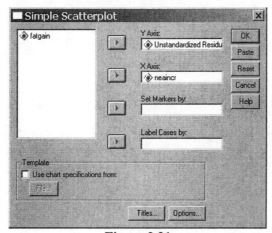

Figure 2.21

4. Click **OK.**
5. When the scatterplot appears in the output window, double-click inside the scatterplot to gain access to the Chart Editor.
6. Once in the Chart Editor, double-click in the graph area of the plot, and then click **Chart, Add Data Element, and Y Axis Reference Line.**
7. The "Properties" box will open.
8. Click on the "Reference Line" tab and here place **"0"** in the **"Y Axis Position"** box. See Figure 2.22 on the following page.

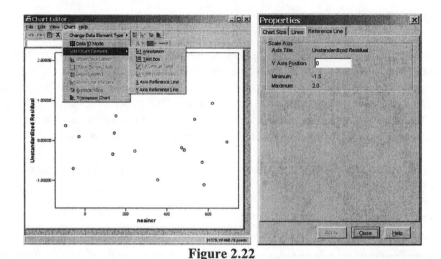

Figure 2.22

The residual plot, with the fit line at $y = 0$, is shown in Figure 2.23. Because there is no pattern to the points in this plot, the fit appears to be satisfactory. Compare this result to Figure 2.20b in IPS.

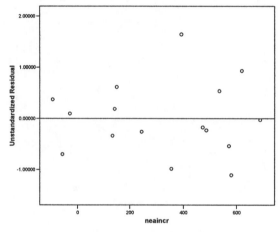

Figure 2.23

Outliers and Influential Observations

An outlier is an observation that lies outside the overall pattern of the other observations. Points that are outliers in the y direction of a scatterplot have large regression residuals making it important to look carefully at the scatterplot and the residual values. Observations that are outliers in the x direction often influence the placement of the regression line. To illustrate we will use the data for Example 2.17 from IPS. The scatterplot for these data is shown in Figure 2.24 on the facing page. The symbols representing Subject 15, an outlier on the y axis, and Subject 18, an outlier on the x axis, have been filled to make them stand out. Compare this figure to Figure 2.22 in IPS.

The residuals have been plotted in Figure 2.25. Compare this outcome to Figure 2.23 in IPS. Notice that Subject 15 has a large residual while Subject 18 does not. To further assess the effects of these two subjects on our least-squares line, we can recalculate the line leaving out the points that are a concern. The comparisons are shown in Figure 2.26. The axes have been scaled the same way in each portion of the figure for comparison. The left hand figure is the original scatterplot. The center figure shows the result of omitting Subject 18 and the right hand figure shows the impact of omitting Subject 15. Compare these plots to Figure 2.24 in IPS.

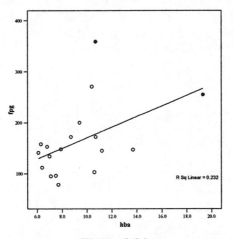

Figure 2.24

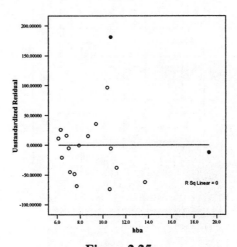

Figure 2.25

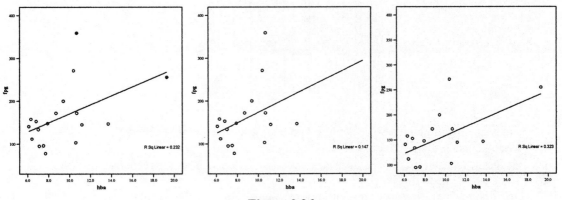

Figure 2.26

Exercises For Chapter 2

1. Three small data sets (X and Y pairs) are shown below. For each one, create a scatterplot and a residual plot. For each pair of plots, place the fitted line (at 0 for the residual plots) and then describe the relationship and what the residual plot tells us about the relationship. Remember that the X variable goes on the horizontal axis for each of these plots.

X	Y	X	Y	X	Y
1	1	1	1.0	1	1.0
2	3	1	1.5	1	1.5
3	4	1	2.0	1	2.0
4	5	1	2.5	1	2.5
5	5	2	2.0	2	2.5
6	3	2	2.5	2	3.0
		2	3.0	2	3.5
		2	3.5	2	4.0
		3	3.0	3	3.0
		3	3.5	3	3.5
		3	4.0	3	4.0
		3	4.5	3	4.5
		4	4.0	4	3.5
		4	4.5	4	4.0
		4	5.0	4	4.5
		4	5.5	4	5.0
		5	5.0	5	3.5
		5	5.5	5	4.0
		5	6.0	5	4.5
		5	6.5	5	5.0
		6	3.0	6	2.5
		6	6.0	6	3.0
		6	6.5	6	3.5
		6	7.0	6	4.0

2. Use the following data to answer the questions that follow.

X	Y
1	−.3
2	−.6
3	−.9
4	−1.2
5	−1.5
6	−1.8
7	−2.1
8	−2.4
9	−2.7
10	−3.0

a. Make a histogram with a normal curve laid over it for the variable called Y.
b. What is the mean and standard deviation for Y?
c. Make a scatterplot using X as the explanatory variable and Y as the response variable.
d. Add a fitted line to the plot.
e. Describe the plot in terms of direction, form, and strength of the correlation.
f. Plot the residuals. What do they tell us about the relationship?
g. Calculate the values for r and r^2. Explain what these values tell us about the relationship between X and Y.

h. What should your response be if you were asked to predict a Y-value for an X-value of 20?
i. Predict a Y-value for an X-value of 4.5 both "by eye" and using SPSS.

3. There is a known correlation between the voluntary homework problems done by statistics students and their final grades. Some representative data are shown below.

Final Course Grade	Homework Problems
55	30
60	40
50	75
60	75
70	75
70	80
50	80
80	80
90	80
70	85
92	85
85	91
92	92
88	98
90	98
100	100

a. Make a histogram with a normal curve laid over it for each of the variables.
b. What is the mean and standard deviation for each?
c. Make a scatterplot by choosing an appropriate explanatory and response variable.
d. Add a fitted line to the plot.
e. Describe the plot in terms of direction, form, and strength of the correlation.
f. Plot the residuals. What do they tell us about the relationship?
g. Are there any outliers? Explain.
h. Calculate the value for r. What does it tell us about the relationship between Final Course Grade and Study Problems?
i. Calculate the value for r^2. What does it tell us about the relationship between Final Course Grade and Study Problems?
j. What should your response be if you were asked to predict a Final Grade for a student who completed no homework problems?
k. Predict a Final Grade for a student who completed 55 homework problems both "by eye" and using SPSS.
l. Add the mean for Final Grade and the mean for Homework Problems to the scatterplot. Where does this point lie?

4. The 1950's were an era of sexual repression and sexual research was in its infancy. Drs. Comfort and Joy reported the following observations on goldfish in October 1967 in the *Journal of Goldfish Studies* (the experiment was conducted in 1956). A large photograph of a hungry cat was presented for varying lengths of time; the exposure time (ET) was recorded.   Simultaneously, a measurement of sexual arousal (SA) was taken. The scores for 5 goldfish are shown on the following page.

ET	SA
1	2
2	7
4	4
6	5
8	7

a. Make a scatterplot for these data appropriately labeled.
b. Add a fit line to the plot.
c. What is the regression (prediction) equation for predicting SA from ET?
d. Describe the relationship in terms of direction, form, and strength.
e. Are there any outliers? Explain.
f. What is the value of r and r^2 for this relationship? Explain what these values mean in terms of ET and SA.
g. What is the mean and standard deviation for ET and SA?
h. Make a histogram for each of ET and SA and place a normal curve over them. Describe the shape, center, and spread for each variable.
i. Make a plot of the residuals. Does a linear relationship provide a good description of the relationship?
j. Discuss the "restricted range" problem in the context of these data.

5. There is good evidence to suggest that studying for exams will improve GPA. There is similar evidence that students with high GPA's are more likely to succeed at graduate school. Here are some fictional data that you can use to answer the questions that follow.

Study Hours	GPA	Completion of Ph.D.
6	1.82	N
42	2.95	Y
31	2.68	N
50	3.52	Y
45	3.71	Y
31	2.5	N
61	3.20	Y
31	3.08	N
40	3.64	Y
26	4.00	Y
35	3.34	N
10	1.20	N

a. Make a histogram for both numeric variables with the normal curve over each. Describe each in terms of shape, center, and spread.
b. Make a scatterplot of the relationship between Study Hours and GPA. Describe direction, form, and strength for this relationship.
c. Make a scatterplot of the relationship between GPA and completion of Ph.D. or not. Describe direction, form, and strength for this relationship.
d. Go back to your scatterplot for Study Hours and GPA and add a categorical variable for whether or not these individuals completed their Ph.D.
e. Determine the regression equation that describes the relationship between Study Hours and GPA.
f. What is the predicted GPA for someone who studies 25 hours per week?
g. What is the predicted GPA for someone who studies 75 hours per week?
h. What is the predicted GPA for someone who studies 50 hours per week?

6. One of the factors affecting the quality of the red wines of the Bordeaux region is the rainfall during the critical months of August and September. In the table below we have vintage quality ratings and the rainfall levels during August and September for the years 1966 to 1978. The rainfall is measured in millimeters. Quality ratings can range from 1 (very poor) to 20 (exceptionally good).

Year	Rainfall	Quality
1966	75.0	18.0
1967	100.0	15.0
1968	258.5	9.0
1969	236.7	13.0
1970	74.6	18.0
1971	118.3	17.0
1972	153.0	13.0
1973	151.4	15.5
1974	183.2	15.0
1975	216.0	18.5
1976	166.6	17.5
1977	67.7	15.0
1978	51.5	18.5

a. Do an exploratory data analysis for Rainfall and Quality: make a histogram showing the normal curve, identify the mean and standard deviation for each variable, and comment on shape, center, and spread.
b. In what percentage of years was there less than 75.0 mm of rain (use the CDF function from Chapter 1)?
c. Plot the relationship between Rainfall in the critical period and Quality showing the year as a label for each data point in the scatterplot.
d. Describe the relationship in terms of direction, form, and strength.
e. Determine the regression equation and place the fit line on the scatterplot.
f. Make a prediction about Quality for a year in which Rainfall is 120 mm.
g. Does Rainfall cause Quality? Explain.

7. In order to evaluate the effectiveness of advertising expenditures, a national retail firm chose seven marketing regions that were similar in retail potential. Then they varied the advertising expenditures across the seven regions. In the table below are the Expenditures (in millions of dollars) and the resulting Sales (in millions of dollars).

Expenditure	Sales
3.2	20.4
1.8	16.4
4.1	24.0
.8	15.9
6.8	27.3
2.3	18.9
2.7	17.8

a. Do an exploratory data analysis for each of these variables: make a histogram showing the normal curve, identify the mean and standard deviation for each variable, and comment on shape, center, and spread.
b. Plot the relationship between Expenditure and Sales. Describe the relationship in terms of direction, form, and strength.
c. Determine the regression equation and place the fit line on the scatterplot.
d. Plot the residuals. Comment on the shape of these and whether or not there are any outliers.

e. Make a prediction about Sales for a region in which Expenditure is 8 million dollars.
f. Make a prediction about Sales for a region in which Expenditure is 5 million dollars.
g. Does Expenditure cause Sales? Explain.

8. In the exercises at the end of Chapter 1, we looked at data adapted from Dement, W. (1960) on the effect of dream deprivation. The data are reproduced below.

Subject	1. Baseline Dream Time As A Percent Of Sleep Time	2. Number Of Dream Deprivation Nights	3. Number Of Awakenings		4. Percent Dream Time Recovery
			3a. First Night	3b. Last Night	
1	19.5	5	8	14	34.0
2	18.8	7	7	24	34.2
3	19.5	5	11	30	17.8
4	18.6	5	7	23	26.3
5	19.3	5	10	20	29.5
6	20.8	4	13	20	29.0
7	17.9	4	22	30	19.8
8	20.8	3	9	13	*

* Subject dropped out of study before recovery nights.

a. Open your data set from the place that you saved it or re-enter the data.
b. For each variable in the data set, do an Exploratory Data Analysis and comment on shape, center, and spread.
c. What is the correlation between the baseline percent dream time and the percent dream time recovery? Make the appropriate scatterplot with a fitted line and residual plot. Derive the values of r and r^2. Comment on direction, form, and strength of the relationship.
d. What is the correlation between the number of awakenings on the first night and dream time recovery? Make the appropriate scatterplot with a fitted line and residual plot. Derive the values of r and r^2. Comment on direction, form, and strength of the relationship.
e. What is the correlation between the number of awakenings on the last night and dream time recovery? Make the appropriate scatterplot with a fitted line and residual plot. Derive the values of r and r^2. Comment on direction, form, and strength of the relationship.
f. Could there be a lurking variable here? *Hint* – it has to do with the number of dream deprivation nights.

Chapter 3. Producing Data

Topics covered in this chapter:

3.1 First Steps
3.2 Design Of Experiments
   Randomization
3.3 Sampling Design
   Simple Random Samples
3.4 Toward Statistical Inference
   Sampling Variability
   Sampling Distributions
   Beyond The Basics: Capture-Recapture Sampling

Statistical designs for producing data rely on either experiments or sampling. An experiment deliberately imposes some treatment on individuals in order to observe their responses. The idea of sampling is to study a part in order to gain information about the whole.

3.1 First Steps

Exploratory data analysis (EDA) is the first step in data analysis. This step is generally followed by specific assessments of statistical inference. These techniques are described throughout the text.

However, before we can complete any type of analysis, we must gather data, either by conducting experiments ourselves or by consulting the data collected by others. In IPS, the authors provide us with several Web sites for accessing data produced by federal agencies in North America. One of these sites is the Statistics Canada site. At this site we can access, for example, the number of university degrees granted in Canada for the years 1995-2000. These data are organized by field of study and sex. Go to: http://www.statcan.ca/english/Pgdb/healtheduc21.htm for these data. Notice that we are given a summary of the data; that is, the detail is given in summary form.

To discover how these data were collected, often important for understanding the data, go to http://www.statcan.ca/endglish/sdds/3124.htm.

3.2 Design of Experiments

Randomization

The use of chance to divide subjects (experimental units) into groups is called randomization. Sampling can be used to randomly select (randomize) treatment groups in an experiment. In Example 3.6 of IPS, researchers want to know if talking on a hands-free cell phone distracts drivers. The response variable is how quickly the subject responds when the car ahead brakes. A control group of students simply drive, whereas the experimental group are to talk on the cell phone while driving. This experiment has one factor (cell phone use) with two levels. The researchers use 40 students for the experiment and so must divide them into two groups of 20.

For large data sets (2000, for example), it is easy to set up the numbers using Excel. For the example shown below, we need the numbers 1 to 40. Place the number 1 in the first cell of an Excel spreadsheet.

In the column below that, click "=", then click in the column above and "=A1" will appear. Now type in "+1" and then press the "enter" key. The second cell in Column 1 will now show "2". Now click in the cell showing "2" and then copy it. Now paste that cell in as often as you wish, for example, up to Row 40 in Column 1. You now have the numbers 1 to 40 in Column 1 of the Excel spreadsheet. Save these data and then open SPSS. Click **File, Open, Data.** Click in the "Files of Type" box until ".xls" appears. Now open the Excel file, for example, *"OnetoForty.xls."* Now continue with the random selection process that is shown below.

A data file was created using the variable *student* (declared as numeric 8.0) that contained the numbers 1 to 40. To randomly assign the 40 students to one of the two groups, use the following instructions:

1.   Click **Data,** and then click **Select Cases.** The "Select Cases" window in Figure 3.1 appears.

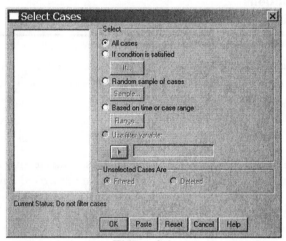

Figure 3.1

2.   Click **Random sample of cases,** and then click the **Sample** button. The "Select Cases: Random Sample" window in Figure 3.2 appears.

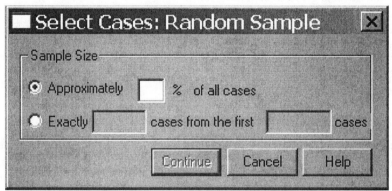

Figure 3.2

3.   Click **Exactly ☐ cases from the first cases,** and fill in the boxes so the line reads "Exactly **20** cases from the first **40** cases."
4.   Click **Continue** and click **OK.**

SPSS for Windows creates a new variable, *filter_$,* which is assigned a value of 0 if the case was not randomly selected and a value of 1 if the case was randomly selected. Furthermore, unselected cases are marked in the Data Editor with an off-diagonal line through the row number, as can be seen in Figure 3.3, which shows the first 10 cases.

	student	filter_$	v
1	1	1	
2	2	1	
3	3	0	
4	4	1	
5	5	1	
6	6	0	
7	7	1	
8	8	1	
9	9	1	
10	10	1	

Figure 3.3

Students 1, 2, 4, 5, 7, 8, 9, 10, 11, 12, 14, 18, 20, 21, 23, 24, 27, 29, 31, and 32 were randomly selected to receive one treatment (the experimental group). The remaining 20 students would then receive the other treatment (the control group). If the *filter_$* variable is deleted and the procedure repeated, a different random assignment of students to the two levels should be obtained. Note that your sample is unlikely to march this one since the sampling process is random.

3.3 Sampling Design

Sampling can also be used to select treatment groups for more complex experimental designs. In the following example, the experimenters investigated the effects of repeated exposure to an advertising message. In this experiment, all subjects will watch a 40-minute television program that includes ads for a digital camera. Subjects will see either a 30-second or 90-second commercial repeated one, three, or five times. This experiment has two factors: length of commercial (with two levels) and number of repetitions (with three levels). The six different combinations of one level of each factor form six treatments.

First, enter the experimental design in an SPSS for Windows Data Editor by listing all the possible combinations and an identifying number for each subject. It requires at least 18 subjects to complete this experiment. The design is shown in Figure 3.4 on the next page.

To randomly assign subjects to treatments, follow the instructions shown above for randomization in experiments choosing 9 out of the 18 subjects. Your outcome will look like Figure 3.5.

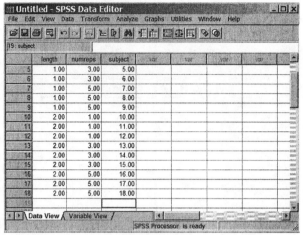

Figure 3.4

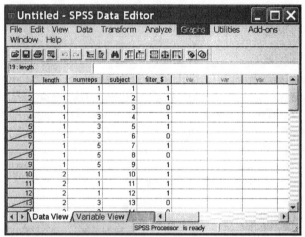

Figure 3.5

Notice that we now need to rearrange our subjects for the experiment. For example, we can assign subjects with a *filter_$* value of 0 to *length* = 1 and those with a *filter_$* value of 1 to *length* = 2. For my example, Subjects 3, 6, 8, 13, 14, 15, 16, 17, and 18 are assigned to *length* = 1 and the remainder to *length* = 2.

Simple Random Samples

SPSS for Windows can be used to **select *n* cases** from a finite population of interest using **simple random sampling.** Simple random sampling, the most basic sampling design, allows impersonal chance to choose the cases for inclusion in the sample, thus eliminating bias in the selection procedure.

Example 3.18 in IPS contains a list of spring break destinations. The authors plan to call a few randomly chosen resorts at each destination to ask about views held towards groups of students as guests. One option for choosing three resorts is to use a table of random digits (Table B). Another option is to use SPSS for Windows to randomly choose 3 of the 28 resorts. A data file was created using *resort* as a variable (declared as a string variable with length 20) that contained the names of the 28 resorts. You can also set this up as a numeric data set using Excel as described above. Table 3.1 on the next page is a reproduction of that list.

01	Aloha Kai	08	Captiva	15	Palm Tree	22	Sea Shell
02	Anchor Down	09	Casa del Mar	16	Radisson	23	Silver Beach
03	Banana Bay	10	Coconuts	17	Ramada	24	Sunset Beach
04	Banyan Tree	11	Diplomat	18	Sandpiper	25	Tradewinds
05	Beach Castle	12	Holiday Inn	19	Sea Castle	26	Tropical Breeze
06	Best Western	13	Lime Tree	20	Sea Club	27	Tropical Shores
07	Cabana	14	Outrigger	21	Sea Grape	28	Veranda

Table 3.1

To select a random sample of *n* cases from *N* cases, follow the steps used earlier in the randomization section above.

1. Click **Data,** and then click **Select Cases.** The "Select Cases" window appears.
2. Click **Random sample of cases,** and then click the **Sample** button. The "Select Cases: Random Sample" window appears.
3. Click **Exactly □ cases from the first cases** and fill in the boxes so the line reads "Exactly **3** cases from the first **28** cases."
4. Click **Continue** and then click **OK.**

The results for the first 10 cases (resorts) are shown in Figure 3.6. With this particular procedure, resorts 09 Casa del Mar and 10 Coconuts are 2 of the 3 cases that were randomly selected to be interviewed. Holiday Inn was the third. If the *filter_$* variable is deleted and the procedure repeated, you should obtain a different set of resorts randomly selected to be interviewed. This is sometimes called sampling with replacement; that is, the original three were "replaced" before the second random sample was selected. Had these first 3 resorts been deleted before the second random sample was completed, we have used "sampling without replacement."

	resort	filter_$	var
1	Aloha Kai	0	
2	Anchor Down	0	
3	Banana Bay	0	
4	Banyan Tree	0	
5	Beach Castle	0	
6	Best Western	0	
7	Cabana	0	
8	Captiva	0	
9	Casa del Mar	1	
10	Coconuts	1	

Figure 3.6

3.4 Toward Statistical Inference

Sampling Variability

Each time that we draw a simple random sample from a population, the outcome will be slightly different. This fact produces what is known as sampling variability. To illustrate, take a second sample of 3 resorts from the set of 28 in Example 3.18. While it is possible that your second sample will be an exact match for your first sample, it is also possible that a completely different set of three resorts will be chosen for your sample. This principle is the basis for sampling variability.

Sampling Distributions

To continue our example of sampling variability and sampling distributions, go to the text Web site at www.whfreeman.com/ips5e and download the WORKERS data set. Using the instructions given earlier, select a simple random sample (SRS) of size 100 from the 14,959 members of this population. Find the average age for the sample. Now repeat this process 10 times and make a histogram of the average age for each of your ten samples. This histogram will give you a visual representation of sampling variability called a sampling distribution. The mean of all ten of these means will give you an estimate of the population mean.

Beyond The Basics: Capture-Recapture Sampling

In some cases, when we complete our second sample, we want all members of the original sample to be included in that second sample. Consider the example of choosing a second set of 3 resorts from a set of 28. Do we leave the original 3 in the data set or remove them before taking the second sample? In our second sample of 3 resorts from a set of 28, we have taken this approach. Note also the cautions about using this approach.

Exercises For Chapter 3

1. Suppose that your university maintains a colony of male mice for research purposes. The ages of the mice are normally distributed with a mean of 60 days and a standard deviation of 5.2. Assume that at any one time there are 2000 mice available.
    a. Set up the numbers 1 to 2000 using Excel. Note: there are ways using the Syntax commands to do all of this in SPSS. The use of Syntax commands is beyond the scope of this manual.
    b. Open the .xls spreadsheet using SPSS.
    c. Generate the sample of ages using the *"rv.normal(mean,stdev)"* to generate the ages of the animals in your simulated population.
    d. Generate a random sample of size 20 for your own research purposes. Use the *"unselected cases are deleted"* option.
    e. For your unique sample, do an exploratory data analysis.
    f. Describe the shape, center, and spread for the ages of the animals in your sample.
    g. Repeat this parts b through f for a second randomly selected set of 20 rats.
    h. Explain why the outcome is slightly different for this second sample in terms of sampling variability.

2. A popular lottery game in Canada is called Lotto 649. Twice each week, a set of six numbers is randomly generated from the set of 49. Generate the next "winning" number.
    a. Set up the numbers 1 through 49 in Excel as shown above.
    b. Open the .xls spreadsheet using SPSS.
    c. Generate a 6-number combination and then sort the numbers in ascending order
    d. Repeat part c for a second randomly selected set of 6 numbers.
    e. Explain why the outcome is slightly different for this second sample in terms of sampling variability.

3. Suppose that Nancy has 25 blouses all color coordinated with her favorite skirts and slacks. She wishes to wear these blouses on 19 different days without repeating any of them. Set up a random selection of 19 blouses for Nancy to wear with her skirts. Now select a second set of 12 blouses for Nancy to wear with her slacks. Explain why Nancy might obtain a different random selection than you did even if she used exactly the same process.

4. Carry out a simulation study with N = 2000 students. Generate their final grades in a statistics class with a mean = 65 and standard deviation = 15.
    a. Generate the sample of grades in your simulated population using the *"rv.normal(mean,stdev)"*
    b. Generate a random sample of size 20 for your own research purposes. Use the *"unselected cases are deleted"* option.
    c. For your unique sample, do an exploratory data analysis.
    d. Describe the shape, center, and spread for the grades of the students in your sample.
    e. Repeat parts b through f for a second randomly selected set of 20 students.
    f. Explain why the outcome is slightly different for this second sample in terms of sampling variability.

5. Suppose that you have an urn with 200 balls in it. Of these 40 are red, 100 are blue, and 60 are green.
    a. Generate a sample containing 8 red ones, 45 blue ones, and 10 green ones.

6. Suppose you wanted to carry out stratified random sampling where there are 3 strata. The first stratum contains 500 elements, the second 400 elements, and the third 100 elements.

a. Generate a stratified sample with 50 elements from the first stratum, 40 elements from the second stratum, and 10 elements from the third stratum.

b. Comment on why this might be called *proportional sampling*.

Chapter 4. Probability: The Study of Randomness

Topics covered in this chapter:

4.1 Randomness
   Summarizing The Results
4.2 Probability Models
   Probability Calculations
4.3 Random Variables

4.1 Randomness

Examples 4.1 and 4.2 in IPS illustrate tossing a coin (two possible outcomes) repeatedly. If the coin is a 'fair' coin, Heads and Tails should turn up equally often "in the long run." We can simulate this process using SPSS for 100 coin tosses by generating a series of 0's and 1's.

1. Using an Excel spreadsheet, as described in the previous chapter, enter the numbers from 1 to 100 in a single column. Save the numbers then open the data set with SPSS.
2. Next choose **Transform** ‣ **Compute** ‣ and the "Compute Variable" window appears.
3. Scroll down the "Functions:" box until ‣ **RV.Bernoulli(p)** appears. Click on this option to move **RV.Bernoulli(?)** into the "Numeric Expression:" box.
4. Click on the **"?"** and change it to *.5* (for equal probabilities of Heads or Tails).   Type a variable name in the "Target Variable:" box (see Figure 4.1).

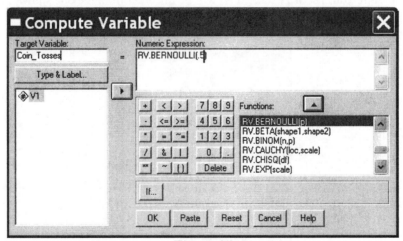

Figure 4.1

5. Click **OK.**
6. Now Click **Transform** ‣ **Recode** ‣ **Into Different Variable.**
7. Click *Coin_Tosses,* then click ‣ to move *Coin_Tosses* into the "Numeric Variable → Output Variable Box."
8. In the "Output Variable" box, type the name for your new variable, and in the "Label:" box elaborate on the name if you would like. Click **Change** (see Figure 4.2).

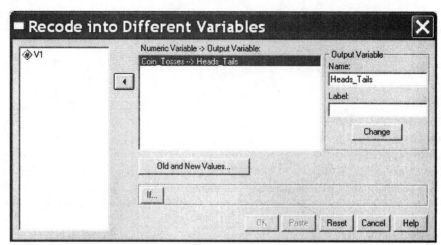

Figure 4.2

9. Click **Old and New Values.**
10. The "Recode into Different Variables: Old and New Values" box will appear. Click on the **"Output Variables are Strings"** box in the lower right corner.
11. In the "Old Value" box type *1.* In the "New Value" box, type *H* then click **Add. 1 → 'H'** appears in the "Old → New box."
12. Repeat this process to recode **0** to **T.**
13. Click **Continue.**
14. Click **OK.**

The results are shown in Figure 4.3. Note that because this is a random generation process, your list of **1**'s and **0**'s (and therefore your list of **H**'s and **T**'s) will be different from the one shown in the figure.

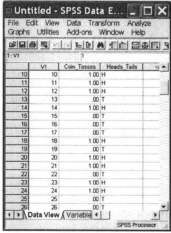

Figure 4.3

Summarizing The Results

To tabulate the results of your 15 coin tosses, follow these steps:

1. Click **Analyze ‣ Descriptives ‣ Frequencies.**
2. Click *Heads_Tails* and ‣ to move *Heads_Tails* into the "Variable(s):" box.

3. Click **Charts** ▸ **Bar graphs** ▸ **Continue** (if you wish a visual description of your outcome in addition to the numerical one).
4. Click **OK.**

The outcome is shown below in Figure 4.4 and Table 4.1.

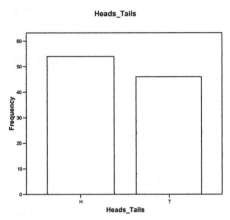

Figure 4.4

Heads_Tails

		Frequency	Percent	Valid Percent	Cumulative Percent
Valid	H	54	54.0	54.0	54.0
	T	46	46.0	46.0	100.0
	Total	100	100.0	100.0	

Table 4.1

As you look at this output, you may be surprised to see that the number of Heads is greater than the number of Tails. Recall that these will be equal only at .5 *in the long run.* To get a more realistic outcome, you need to toss the coin thousands of times. See Example 4.2 in IPS for some suggestions about how often to toss the coin.

4.2 Probability Models

There are several other distributions for which SPSS for Windows can calculate probabilities. Let us first look at binomial probabilities using the following example. An engineer chooses a simple random sample of 10 switches from a shipment of 10,000 switches. Suppose that, unknown to the engineer, 10% of the switches in the shipment are bad. The engineer counts X, the number of bad switches in the sample. What is the probability that no more than 1 of the 10 switches in the sample fails inspection?

Let X = the number of switches in the sample that fail to meet the specifications. X is a binomial random variable with $n = 10$ and $p = .10$. To find the probability that no more than 1 of the 10 switches in the sample fails inspection ($P(X \le 1)$ when $n = 10$ and $p = .10$), follow these steps:

1. Define a new variable (e.g., *Switch*), and enter the values 0 through 10. This variable was declared a numeric variable of length 8.0.
2. Click **Transform** and then click **Compute.** The "Compute Variable" window in Figure 4.5 appears.

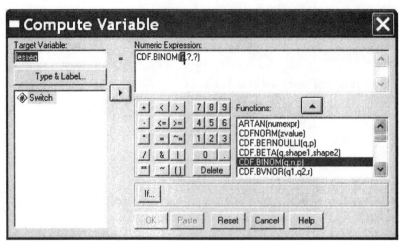

Figure 4.5

3. In the "Target Variable:" box, type in *lesseq.*
4. In the "Functions" box, click ▾ until **CDF.BINOM(q,n,p)** appears in the box. Double-click on **CDF.BINOM(q,n,p)** to move **CDF.BINOM(?,?,?)** into the "Numeric Expression:" box. The **CDF.BINOM(q,n,p)** function stands for the cumulative distribution function for the binomial distribution, and it calculates the cumulative probability that the variable takes a value <u>less than or equal to</u> *q*.
5. In the "Numeric Expression:" box, highlight the first **?,** click on *Switch,* and click ▸ so *Switch* replaces the first **?.** Replace the second **?** with the number **10** (the value of *n*), then replace the third **?** with the value **.10** (the value of *p*).
6. Click **OK.**

Figure 4.6 displays the new variable *lesseq.* By default, SPSS shows the values of *lesseq* to two decimal places of accuracy. The number of decimal places was changed to four. For each value of *Switch,* the variable *lesseq* represents the cumulative probability of observing that number or fewer failures. We want to determine the probability that no more than 1 of the 10 switches in the sample fails inspection or, symbolically, $P(X \leq 1 \mid n = 10, p = .10)$. The variable *lesseq* tells us that the probability that no more than 1 of the 10 switches in the sample fails inspection is 0.7361, given that the probability that any given switch will fail is $p = .10$.

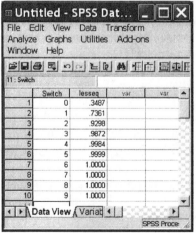

Figure 4.6

Given the same values of n and p, suppose one wants to find the probability that strictly less than two switches fail inspection, $P(X < 2 \mid n = 10, p = .10)$. Table 4.2 gives us the required SPSS format. Steps 2 to 6 can be repeated, with the changes that the target variable is named ***less*** and the "Numeric Expression:" box reads **CDF.BINOM(switch-1,10,.10).** In the output, follow across the row for n=2 to get the probability that less than 2 switches fail.

If one wants to determine the probability that exactly two of the 10 switches fail inspection, $P(X = 2 \mid n = 10, p = .10)$, steps 2 to 6 can be repeated, with the changes that the target variable is named ***equal*** and the "Numeric Expression:" box reads **CDF.BINOM(switch,10,.10) - CDF.BINOM(switch-1,10,.10).**

To determine the probability that at least two of the 10 switches fail inspection, $P(X \geq 2 \mid n = 10, p = .10)$, steps 2 to 6 can be repeated, with the changes that the target variable is named ***greateq*** and the "Numeric Expression:" box reads **1 - CDF.BINOM(switch-1,10,.10).**

Finally, if one wants to determine the probability that no more than two of the 10 switches fail inspection, $P(X > 2 \mid n = 10, p = .10)$, steps 2 to 6 can be repeated, with the changes that the target variable is named ***greater*** and the "Numeric Expression:" box reads **CDF.BINOM(switch,10,.10).** The probabilities associated with the first three switches are shown in Figure 4.7. More specifically, Row 3 shows the probabilities associated with $X \leq 2$, $X < 2$, $X = 2$, $X \geq 2$, and $X > 2$ under the variables of ***lesseq, less, equal, greateq,*** and ***greater,*** respectively.

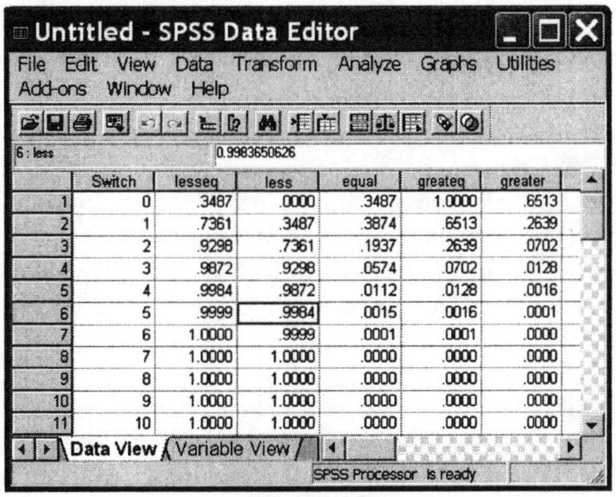

	Switch	lesseq	less	equal	greateq	greater
1	0	.3487	.0000	.3487	1.0000	.6513
2	1	.7361	.3487	.3874	.6513	.2639
3	2	.9298	.7361	.1937	.2639	.0702
4	3	.9872	.9298	.0574	.0702	.0128
5	4	.9984	.9872	.0112	.0128	.0016
6	5	.9999	.9984	.0015	.0016	.0001
7	6	1.0000	.9999	.0001	.0001	.0000
8	7	1.0000	1.0000	.0000	.0000	.0000
9	8	1.0000	1.0000	.0000	.0000	.0000
10	9	1.0000	1.0000	.0000	.0000	.0000
11	10	1.0000	1.0000	.0000	.0000	.0000

Figure 4.7

The formats needed to obtain these various binomial probabilities are summarized in Table 4.2. The symbol x is used to represent some number (of switches). For instance, $P(X < x)$ could represent the expression $P(X < 5$ switches) or the expression $P(X \geq 12$ people who agree).

Binomial Probability Calculations

Binomial Probability Sought	CDF. BINOM Format Needed
$P(X < x)$	CDF.BINOM(x – 1,n,p)
$P(X \leq x)$	CDF.BINOM(x,n,p)
$P(X = x)$	CDF.BINOM(x,n,p) – CDF.BINOM(x – 1,n,p)
$P(X \geq x)$	1 – CDF.BINOM(x – 1,n,p)
$P(X > x)$	1 – CDF.BINOM(x,n,p)

Table 4.2

Here is another example. An opinion poll asks 2500 adults whether they agree or disagree that "I like buying new clothes, but shopping is often frustrating and time consuming." Suppose that 60% of all U.S. residents would say "Agree."

To find the probability that at least 1520 adults agree, that is $P(X \geq 1520)$, create a variable named **_agree_** and enter the single value **_1520_**. Use the steps outlined above until the "Compute Variable" Window appears. Give your new variable a name and insert the following in the "Numeric Expression:" box: **1-CDF.BINOM(agree-1,2500,.6).**

SPSS returns the value of .2131, meaning that the probability that 1520 people will agree is .21.

SPSS for Windows is capable of computing probabilities for a number of additional distributions. Table 4.3 displays a number of commonly used distributions and their commands.

Distribution	SPSS Command
Chi-square	CDF.CHISQ(q,df)
Exponential	CDF.EXP(q,scale)
F	CDF.F(q,df1,df2)
Geometric	CDF.GEOM(q,p)
Hypergeometric	CDF.HYPER(q,total,sample,hits)
Normal	CDF.NORMAL(q,mean,stddev)
Poisson	CDF.POISSON(q,mean)
Uniform	CDF.UNIFORM(q,min,max)

Table 4.3

Probability Calculations

There is a calculator within SPSS that can be used to do arithmetic calculation such as those needed for the basic rules of probability. The calculator performs basic arithmetic operations, such as addition (+), subtraction (—), multiplication (*), division (/), and exponentiation (**). These operations can be used for calculations based on the probability rules described in IPS. You can type the function in or choose it from the function box. The "Compute Variable" window can be accessed by clicking **Transform** then **Compute.**

In order to use the calculator, it is necessary to have some data in the SPSS spreadsheet. To illustrate the process, consider the following example. During WWII, for each British bomber mission the probability of losing the bomber was .05. The probability that the bomber returned therefore was $1 - .05 = .95$. It seems reasonable to assume that missions are independent. Therefore, the probability of surviving 20 missions is:

$$P(A_1 \text{ and } A_2 \text{ and } \ldots \text{ and } A_{20}) = P(A_1)*P(A_2)* \ldots *P(A_{20}) = (.95)^{20}$$

To calculate this probability, first enter the number .95 in the first cell of an SPSS spreadsheet and then complete the following steps:

1. Click **Transform** then **Compute,** and in the "Compute Variable" window, type in the name of the "Target Variable:"
2. Click on the existing variable, then click ▸ so that the existing variable moves into the "Numeric Expression:" window.
3. Following the variable name in the "Numeric Expression:" window, type in **_**20_** (meaning raised to the power of 20 or multiplied by itself 20 times) (see Figure 4.8), then click **OK.**

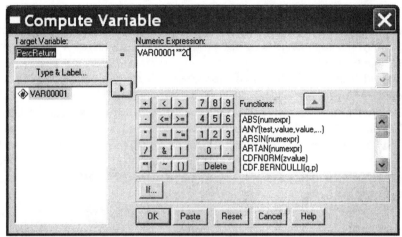

Figure 4.8

The result, shown in Figure 4.9, is that the probability of returning from 20 missions is approximately .36.

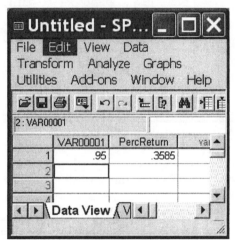

Figure 4.9

This calculator can be used for any of the expressions used to compute conditional probabilities and addition and multiplication rules.

4.3 Random Variables

To this point in the chapter, we have dealt with simulating distributions when we know something about the probability of an event. What about situations in which we know only the shape, center, and spread for our population and wish to generate a sample?

In most universities, the final grades in courses that are fairly large are normally distributed with a mean of 50 and a standard deviation of 15. Such a distribution makes the dean and registrar quite happy. Any grade distribution that is skewed toward either the high or the low end will be a cause for concern. The definition of a large class is not clearly specified, but a class of 50 has been chosen for this example.

To generate several normally distributed random samples for classes of size 50 with a mean of 50 and a standard deviation of 15, follow the steps shown on the next page.

1. First set up the numbers 1 to 50 in a column using Excel and save the spreadsheet. Note: there are ways of using the Syntax commands in SPSS to do this. The use of Syntax commands will be introduced later in this manual.
2. Open the .xls spreadsheet using SPSS and name the variable *start.*
3. Click **Transform** ‣ **Compute** ‣ and the "Compute Variable" window appears.
4. Scroll down the "Functions:" box until ‣ **RV.Normal(mean,stddev)** appears. Click on this option to move **RV.Normal (?,?)** into the "Numeric Expression:" box.
5. Click on the first **"?"** and change it to *50* (for the mean), and then click on the second **"?"** and change it to *15* (for the standard deviation).   Type a variable name in the "Target Variable:" box.   The variable name *Out1* is used in this example (see Figure 4.10). Click **OK.**

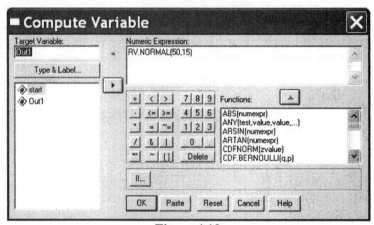

Figure 4.10

6. Repeat this process until you have 10 samples.
7. Calculate **Descriptives** for *Out1* to *Out10* (see Table 4.4).

Descriptive Statistics

	N	Minimum	Maximum	Mean	Std. Deviation
Out1	50	14.11	85.20	46.9564	17.00536
Out2	50	4.05	76.13	48.6288	15.29747
Out3	50	21.96	85.42	50.3095	15.91811
Out4	50	30.31	81.98	54.7445	11.62751
Out5	50	26.35	85.46	52.0459	15.54856
Out6	50	13.32	72.62	49.2024	12.01058
Out7	50	27.33	88.77	53.3154	13.45983
Out8	50	18.66	78.96	49.2924	13.72692
Out9	50	4.78	87.88	49.6114	16.53140
Out10	50	20.09	88.46	50.3076	14.10340
Valid N (listwise)	50				

Table 4.4

Notice that the minimum and maximum scores are quite different from sample to sample; however, the mean centers around 50 for all 10 samples and the standard deviation centers around 15. The dean and registrar of your faculty would be pleased with these 10 sets of grades. Note that you can see the individual grades of each "class member" by scanning the SPSS for Windows Data Editor content that you generated. Also, because these are random samples, your outcome will be different from that shown here and from that of your classmates. However, overall, you can expect to have a mean of approximately 50 and a standard deviation of approximately 15 for each sample that is randomly generated.

You can use several other types of distributions to generate random samples when specific characteristics of the population are known. More on this in later chapters of the manual.

The final two sections of the chapter are important for our understanding of the properties of variables. There are some calculations involved and you may choose to use your calculator or the **Transform, Compute** functions in SPSS. A thorough understanding of these sections is important. Take time to work through them before continuing on to Chapter 5.

Exercises For Chapter 4

1. What is the probability for each of the following events assuming that in each case you are starting with a complete set of 52 cards in a deck of playing cards:
    a. Drawing an ace of spades?
    b. Drawing any red card?
    c. Drawing anything but the ace of hearts?
    d. Drawing any heart?
    e. Drawing the queen of hearts?

2. In a container there are 10 balls: 2 white, 2 red, 5 blue, and 1 black. Determine the probability for each of the following outcomes:
    a. Not white or blue
    b. Blue or red
    c. Black, white, or green
    d. A red and then a black with no replacement
    e. A red and then a blue with no replacement
    f. A black and then a black with no replacement
    g. The sequence: blue, red, white, blue, black with no replacement

3. Consider a slot machine with three independent wheels (a fair machine). Each wheel contains pictures of fruit that line up to be seen through a window. Suppose you are playing a machine that has pictures of seven different kinds of fruit on each wheel. What is the probability of:
    a. 3 lemons?
    b. no lemons?
    c. exactly 1 lemon?
    d. 1 lemon and 2 cherries?
    e. 3 apples?
    f. 1 apple, 1 lemon, 1 cherry?

4. As you graduate, suppose that you apply for two jobs. You estimate the probability of obtaining the first job at .40 and the probability of being hired for the second job to be .20. Assuming independence:
    a. What is the probability of obtaining either position?
    b. If you apply for a third position and estimate the probability of obtaining the third position to be .10, what is the probability of being offered all three positions?

5. In a seven-horse race, what is the probability of correctly predicting, by chance:
    a. The order of finish of the entire field?
    b. The top three horses without regard to order?
    c. The first and second place horses, in order?

6. Suppose a population contains 50% males and 10% left-handers. The percentage of people who are male and left-handed is 52%.
    a. What percentage is male and left-handed?
    b. Given that a person is female, what is the probability that she is left-handed?

7. An urn contains 10 balls numbered 1 to 10. In drawing with replacement, what is the probability that the ball numbered 1 will be drawn 3 out of 6 draws?

Chapter 5. Sampling Distributions

Topics covered in this chapter:

Chapter 4 of IPS introduced the concept of cumulative density functions. This chapter explores further examples with various types of distributions.

5.1 Sampling Distributions For Counts And Proportions

Finding Binomial Probabilities: Software And Tables

Examples 5.3 and 5.4 in IPS, report some information about tax audits. Individuals are often chosen at random for such an audit. In the example we are given the information that a Simple Random Sample (SRS) of 150/10,000 records is selected. After reviewing these tax records, the auditors find that 800/10,000 (p = .08) are incorrectly classified as subject to sales tax or not.

Let X = the number of misclassified records. X is a binomial random variable with n = 150 and p = 0.08. To find the probability that exactly 4 of the 150 records are misclassified ($P(X = 4)$ when n = 150 and p = 0.08), follow these steps:

1. Define a new variable (e.g., *misclass*), which takes on the values 0 through 149. This variable was declared a numeric variable of length 8.0. You may wish to refer to earlier chapters of this manual for instructions for doing this.
2. Click **Transform** and then click **Compute**. In the "Target Variable" box, type in *equal.*
3. In the "Functions" box, click ▼ until **CDF.BINOM(q,n,p)** appears in the box. Double-click on **CDF.BINOM(q,n,p)** to move **CDF.BINOM(?,?,?)** into the "Numeric Expression" box. The **CDF.BINOM(q,n,p)** function stands for the cumulative distribution function for the binomial distribution, and it calculates the cumulative probability that the variable takes a value <u>less than or equal to</u> q. For more information about each of these functions, right click once you have highlighted the function and an explanatory box will appear like the one in Figure 5.1on the next page.
4. In the "Numeric Expression" box, highlight the first *?,* click on *misclass,* and click ▸ so *misclass* replaces the first *?*. Replace the second *?* with the number **150** (the value of n), then replace the third *?* with the value **0.08** (the value of p).
5. Click **OK.**

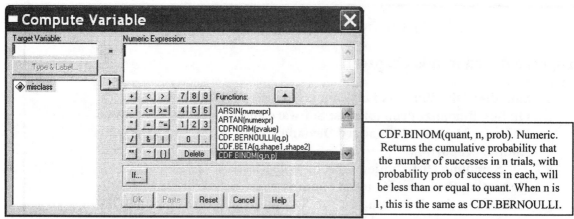

Figure 5.1

Figure 5.2 displays the new variables ***equal*** and ***lesseq.*** By default, SPSS shows these values to two decimal places of accuracy. The number of decimal places here was changed to six.

	misclass	equal	lesseq	var	var
4	3	.001341	.001706		
5	4	.004286	.005991		
6	5	.010882	.016874		
7	6	.022868	.039742		
8	7	.040907	.080649		
9	8	.063584	.144233		
10	9	.087236	.231469		
11	10	.106959	.338427		
12	11	.118373	.456800		
13	12	.119231	.576032		

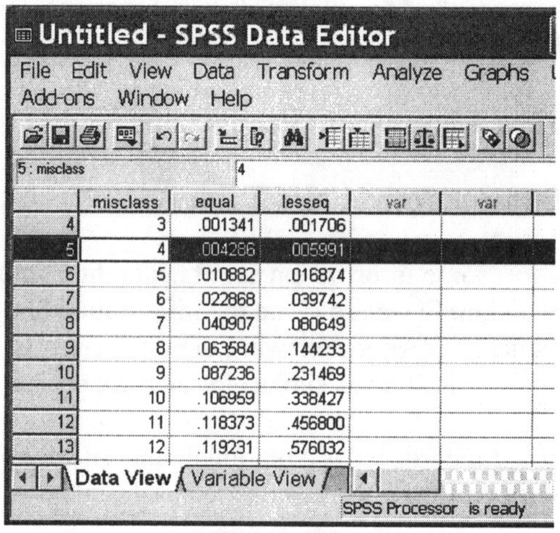

Figure 5.2

For each value of ***misclass,*** the variable ***equal*** represents the cumulative probability of observing that number of misclassifications and the variable ***lesseq*** gives us the cumulative probability of observing that number or fewer misclassifications.

Binomial Mean And Standard Deviation

You can use the "Compute" function in SPSS to make the calculations shown in Example 5.7 or use your desk calculator.

Sample Proportions

You can use the "Compute" function in SPSS to make the calculations shown in Example 5.8 or use your desk calculator.

5.2 The Sampling Distribution Of A Sample Mean

Simulating Repeated Random Samples

A simulation can be used to demonstrate how a confidence interval works:

1. Open your internet browser and go to: www://bcs.whfreeman.com/ips5e/
2. Under **Student Resources,** click on **Statistical Applets.** Then click on **Confidence Interval.**
3. Follow the instructions on the screen to generate a picture like the one in IPS Figure 13.4.

What happens as you add one sample at a time? Fifty samples at a time? What does this visual representation tell you about sampling *in the long run*? What Theorem does this representation illustrate?

Generating Random Samples For Normal Distributions Using SPSS

Psychologists define IQ as the ratio of mental age to chronological age multiplied by 100. For someone whose mental age and chronological age match, they are said to have an IQ of 100. If one's mental age is greater than chronological age, IQ will be above 100. For one common measure of IQ, the standard deviation is 15. Generate a random sample of size 150 with a normal distribution and a standard deviation of 15. What is the mean and standard error (standard deviation divided by the square root of n) of this sample? To complete this question, follow these steps:

1. Enter the numbers 1 to 150 in a column labeled *start*.
2. Click on **Transform, Compute.**
3. Enter the variable name IQ1 in the "Target Variable:" box.
4. Double-lick on **RV.NORMAL(mean, stdev)** in the "Functions: box
5. Replace the first **?** with the mean (100) and the second **?** with the standard deviation (15).
6. Click **OK.** See Figures 5.3 and 5.4.
7. Repeat this process 10 times. See Table 5.1 on the next page.

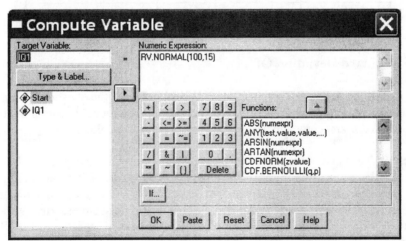

Figure 5.3

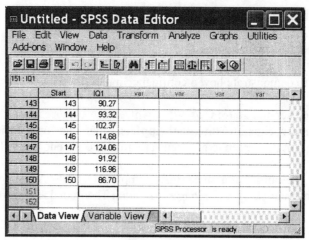

Figure 5.4

Descriptive Statistics

	N	Minimum	Maximum	Mean	Std. Deviation
IQ1	150	54.05	135.20	97.9739	15.23275
IQ2	150	71.96	135.46	102.3666	14.51365
IQ3	150	63.32	138.77	100.6034	13.14106
IQ4	150	54.78	138.46	99.8970	14.37576
IQ5	150	69.34	139.34	99.4786	14.16960
IQ6	150	62.13	139.63	100.6157	14.75661
IQ7	150	64.05	141.61	102.0965	14.43629
IQ8	150	61.68	134.68	99.3709	15.05136
IQ9	150	61.59	136.13	101.2304	14.60330
IQ10	150	58.10	145.74	100.6263	16.19126
Valid N (listwise)	150				

Table 5.1

The Mean And Standard Deviation Of $\bar{x}$

Now we have 10 means, one for each of our samples. We can now ask the question "What happens to the mean over repeated random samples?" To answer this question, create a new column in your SPSS Data Editor window called *XBar* and copy and paste the 10 means into that column.

Now complete descriptive statistics for the new variable *XBar*. This will give you an idea of how the mean varies over repeated random sampling. The output can be seen as a histogram in Figure 5.5. Also, look at Table 5.2. Here we see the mean and standard deviation of the means for our 10 samples. Notice that the calculations duplicate what we have done in the past for single samples. Here we are looking at the center and spread of the means of repeated samples. What is labeled by SPSS as the "Std. Deviation" of our variable *XBar* is actually the standard error of the mean referenced in IPS.

This procedure replicates the process outlined for Example 5.15 in IPS. Notice that one of our samples had a relatively low mean (97.97) and that this has given our normal curve some skew to the left. Overall the mean of all of the means is 100.43 which is very close to the mean of 100 specified for the population. Notice also that not all of the samples have a mean of exactly 100. This is another example of sampling variability.

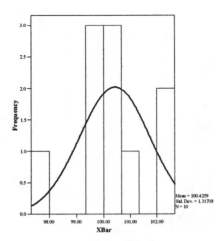

Figure 5.5

Descriptive Statistics

	N	Minimum	Maximum	Mean	Std. Deviation
XBar	10	97.97	102.37	100.4259	1.31718
Valid N (listwise)	10				

Table 5.2

The Central Limit Theorem

Example 5.17 in IPS is a detailed example relating to the central limit theorem. For this example, the distribution is strongly right-skewed, and the most probable outcomes are near 0. This distribution, called an exponential distribution and has a mean $\mu = 1$ and a standard deviation $\sigma = 1$. Figure 5.10 in IPS illustrates the central limit theorem for this distribution. The following steps outline how to use SPSS to illustrate the central limit theorem.

First, we need to generate a large number of large samples. The definitions of large in this context will be set arbitrarily as 25 random samples of size n = 250. To generate 25 exponentially distributed random samples for samples of size 250 with a mean of 1 and a standard deviation of 1, follow these steps:

1. First, we need a "starting" number. In the first column of your data set type in a number of your choosing and repeat the number until it occurs 250 times in that column. I chose to do this using an Excel spreadsheet and called the file N=250. Then I opened the file using SPSS. The number 37 was chosen for this example and the variable was named *start.*
2. Click **Transform** ▸ **Compute** ▸ and the "Compute Variable" window appears.
3. Scroll down the "Functions:" box until ▸ **RV.EXP(scale)** appears. Click on this option to move **RV.EXP(?)** into the "Numeric Expression:" box.
4. Click on the "?" and change it to *1.* Type a variable name in the "Target Variable:" box. In this example, the variable name ***population*** was used.
5. Click **OK.** The distribution for our population is shown in Figure 5.6. Compare this to Figure 5.10(a) in IPS.
6. Now take multiple random samples from this population with sample sizes of 2, 10 and 25.
7. Follow the instructions given earlier in the chapter and create at least 10 samples for each sample size, calculate and plot the mean of these means. See Figure 5.7. Compare these to Figures 5.10 (b), (c), and (d) in IPS.

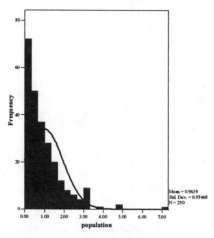

Figure 5.6.

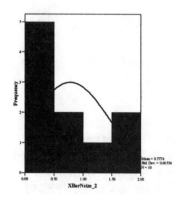 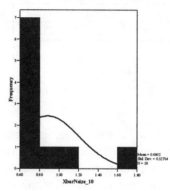 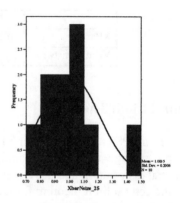

Figure 5.7

Other SPSS Functions

Shown below are some examples, taken from the SPSS help file for random variables (RV). Pressing the Help key while in the "Compute Variables:" window will take you to these and other options.

The following functions give a random variate from a specified distribution. The arguments are the parameters of the distribution. You can repeat the sequence of pseudo-random numbers by setting a seed in the Preferences dialog box before each sequence. Note the period in each function name.

NORMAL(stddev) Numeric. Returns a normally distributed pseudo-random number from a distribution with mean 0 and standard deviation stddev, which must be a positive number. You can repeat the sequence of pseudo-random numbers by setting a seed in the Preferences dialog box before each sequence.

RV.BERNOULLI(prob) Numeric. Returns a random value from a Bernoulli distribution with the specified probability parameter prob.

RV.BINOM(n, prob) Numeric. Returns a random value from a binomial distribution with the specified number of trials and probability parameter.

RV.CHISQ(df) Numeric. Returns a random value from a chi-square distribution with the specified degrees of freedom df.

RV.EXP(shape) Numeric. Returns a random value from an exponential distribution with the specified shape parameter.

RV.LNORMAL(a, b) Numeric. Returns a random value from a log-normal distribution with the specified parameters.

RV.NORMAL(mean, stddev) Numeric. Returns a random value from a normal distribution with the specified mean and standard deviation.

RV.UNIFORM(min, max) Numeric. Returns a random value from a uniform distribution with the specified minimum and maximum. Also see the UNIFORM function.

Exercises For Chapter 5

1) An SRS of 100/5000 students is selected to answer the question: "Do you approve of the sale of lottery tickets on campus or not?" Of these students, 40% disapprove.
   a) Find the probability that exactly 30 of the students disapprove.
   b) Find the probability that more than 70 disapprove.
   c) Suppose that only 25 students were sampled. What is the probability that no more than 1 of these would disapprove?
   d) Are these independent observations?
   e) What is the count of "successes" in this example?
   f) Find the mean and standard deviation of this binomial distribution.

2) Suppose that we take a random sample of the adult population of size 2000 and ask if individuals take a daily multivitamin or not. Suppose that 55% of our subjects would say yes if asked this question.
   a) Find the mean and standard deviation of the sample proportion.
   b) What is the probability that the sample proportion who agree is at least 45%?
   c) Explain why we need such a large sample in this situation.
   d) If this sampling process was repeated hundreds of times, what would the distribution of sample proportions look like? Why?

3) * OPTIONAL Each child born to a set of parents has a probability of .5 of being male or female. A particular set of parents wishes to have all boys. Suppose these parents have 5 children.
   a) What is the probability that exactly 2 of them are girls?
   b) What is the probability that 1 or more is a boy?
   c) What is the probability that all 5 children are boys?

4) Generate an exponential curve with a standard deviation of 5 and a sample size of 150.
   a) Plot a distribution for your data set with a normal curve superimposed. Describe the shape of the data. Describe a situation in which you might expect a distribution like this.
   b) Take 10 samples of size 50 from this data set.
   c) Compute the mean of each sample.
   d) Plot this set of 10 means and do EDA on them (the means).
   e) Describe the shape of the sampling distribution of the mean.
   f) What does this tell you about the overall average of the exponential data set?
   g) Relate this outcome to the central limit theorem.
   h) What would the sampling distribution of the mean have looked like if 10 samples of size 2 instead of size 50 were taken?

5) Suppose that a computer technician generally repairs a computer in 30 minutes. The standard deviation is 24 minutes. Repair times generally are exponential distributions. If a technician repairs 100 computers per week:
   a) What is the probability that the average maintenance time exceeds 45 minutes?
   b) Calculate the mean and standard deviation for this sample based on the central limit theorem.

6) * OPTIONAL Weibull curves are common in manufacturing processes. For the intrepid explorer, SPSS will generate Weibull curves given the expected probability and the failure count. Using a sample size of 150:
   a) Generate a Weibull (RV.WEIBULL) curve with the parameters .95 and 1. Compare the histogram of the output to the center curve in Figure 5.14 of IPS.

b) Generate a Weibull curve with the parameters .1 and 145. Compare this to the lower curve in Figure 5.14 of IPS.

c) Write a description of situations in which each of these curves might apply.

Chapter 6. Introduction to Inference

Topics covered in this chapter:

6.1 Estimating With Confidence

A **confidence interval** is a procedure for estimating a population parameter using observed data or statistics. This chapter in IPS describes the reasoning used in statistical inference and introduces us to cases that require that we know the population standard deviation, σ. This section describes how to examine various **properties of confidence intervals.**

Confidence Intervals

Assume that you are given IQ scores for 31 seventh-grade girls in a Midwest school district. We want to verify that there are no major departures from normality and to obtain the 99% confidence interval for the IQ of the population of seventh-grade girls. We are given $\sigma = 15$.

To calculate the confidence interval using SPSS, follow these steps:

1. Click **Analyze ▸ Descriptive Statistics ▸ Explore.**
2. Click on *iq* and then ▸ to move *iq* into the "Dependent List" window.
3. Click "Plots" in the lower right corner, and then click on "Normality Plots with Tests."
4. Click **Continue** and then click **OK.**

To verify that there are no major departures from normality, check the stem-and-leaf for outliers and then look at the Normal Q-Q Plot. The Normal Q-Q plot for these *iq* data is shown in Figure 6.1. If all the points are near the straight line, as they are in this example, we have good evidence for a normal distribution in this type of visual presentation.

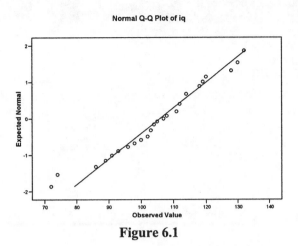

Figure 6.1

Information required to compute the 95% confidence interval can be found in the Descriptives output shown in Table 6.1. The confidence interval noted in the Descriptives table is for $\bar{x}$.

Descriptives

		Statistic	Std. Error
Mean		105.84	2.563
95% Confidence Interval for Mean	Lower Bound	100.60	
	Upper Bound	111.07	
5% Trimmed Mean		106.27	
Median		107.00	
Variance		203.673	
Std. Deviation		14.271	
Minimum		72	
Maximum		132	
Range		60	
Interquartile Range		16	
Skewness		-.470	.421
Kurtosis		.431	.821

Table 6.1

Confidence Interval For A Population Mean

Margin Of Error

The illustrations in this segment of the chapter are based on Example 6.3 in IPS. The calculations can be completed on your calculator or using SPSS. To compute the margin of error, *m*, follow the steps outlined below.

1. To compute *m*, enter the n, σ, and value(s) of z^* into a spreadsheet. Use the **Transform ▸ Compute** features and the formulas given in IPS. See Figure 6.2 for an example.
2. Click **OK** and the value 2684 is returned to the data spreadsheet under the variable labeled *m*.

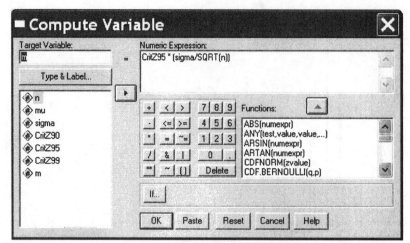

Figure 6.2

This value, when rounded to the nearest $100, is then used to calculate the 95% confidence interval. Again, you can use SPSS to do this or use your calculator. To calculate the upper bound of the confidence interval, add m to $\bar{x}$. To calculate the lower confidence limit, make the subtraction.

What happens to the confidence interval as the sample size is changed? In Example 6.4 in IPS, we have the same population parameters but a sample size of 320 instead of 1280. When we complete the calculations, as above, with the new sample size, our value of **m** becomes 5369 or 5400 when rounded to the nearest $100. Notice that **m** is now twice as large as when we used a sample size of 1280. Now when you calculate the upper and lower bounds of the 95% confidence interval, you will see that it is also larger (a wider interval). This is precisely what happens with smaller sample sizes.

Using SPSS it is easy to demonstrate what happens to the confidence interval as the values for $z*$, n, and σ are changed. Example 6.5 in IPS demonstrates the use of a 99% confidence interval and how **m** is affected. The calculations using SPSS can be replicated as above.

First, change the sample size back to the original value of 1280. Now complete the calculations using $z* = 2.576$. Figure 6.3 shows the outcome.

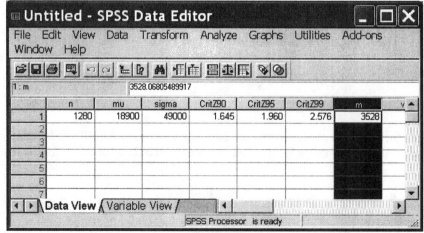

Figure 6.3

Choosing The Sample Size

By choosing your sample size before you collect and analyze data, you can arrange to have both high confidence and a small margin of error. Suppose that we wish a margin of error *m* and to be 95% confident that our confidence interval contains the actual population mean. This situation is described in Example 6.6 in IPS. Again, we can use the Compute function in SPSS to make the calculations. Enter the equation for n in the "Numeric Expression" window as shown in Figure 6.4. The "**" in the equation is our way of telling SPSS to square the result of the calculations inside the brackets. SPSS returns the value of 2305.6 meaning that to achieve a margin of error of $2000 with 95% confidence, we need a sample size of 2306.

When we replicate the steps shown above for a margin of error of $1500, we see that we need almost twice as many subjects; n = 4099.1.

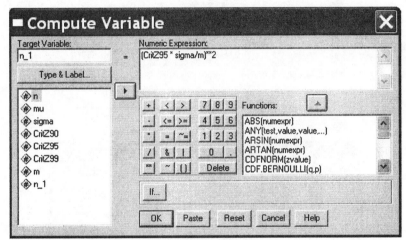

Figure 6.4

Beyond The Basics: The Bootstrap

Confidence intervals are based on sampling distributions. We now know the basic characteristics of a normal distribution. When the data are an SRS from a large population of size N, with mean equal to μ and standard deviation equal to σ, we can find the sampling distribution for $\bar{x}$.

When we do not have a normal distribution and we have a small sample size, what are our alternatives to the sampling distribution calculations already discussed? Bootstrap procedures are a means of approximating sampling distributions when we do not know, or cannot predict using theoretical approaches, their shape. The bootstrap is practical only when you can use a computer to take 1000 or more samples quickly. It is an example of how the use of fast and easy computing is changing the way we do statistics. More details about the bootstrap can be found in a later chapter of IPS.

For a simple illustration in SPSS, enter the set of four weights given in IPS: 190.5 189.0 195.5 187.0. Now follow these steps to get one sample with replacement. Note that each time you do this you are likely to get a slightly different sample.

1. Click **Data**, **Select Cases** and the screen shown in Figure 6.5 appears.
2. Click the button beside "Random sample of cases" and then click "Sample" to move to the screen shown in Figure 6.6.

3. Click on "Exactly" and then specify 1 of the 4 cases that are in the data set (always use the whole data set).
4. Click **Continue.** Click **OK.**

The results are shown in Figure 6.7.

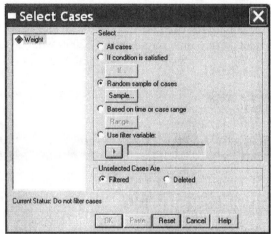

Figure 6.5

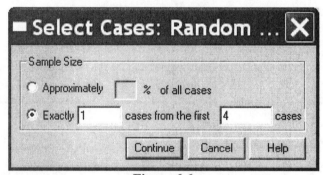

Figure 6.6

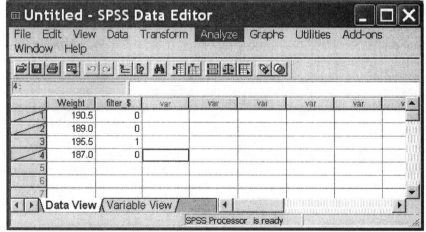

Figure 6.7

Notice that SPSS has created a filter variable for us that returns 1's and 0's. As requested, there is a "1" beside only one of our four cases. This now becomes the first value in our resample. Copy this value and paste it into the first cell of an empty column and label the column *Resample_1*. Repeat this process until you have a new sample of size 4 in *Resample_1*. See Figure 6.8 for one possible outcome.

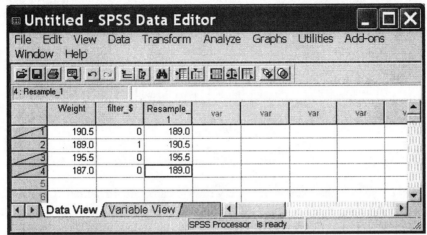

Figure 6.8

Now, we need to repeat this entire process of sampling 1 case at a time, with replacement, from our four cases until we have approximately 1000 resamples. This is not really practical using the method outlined here. We will discuss this procedure more fully in a later chapter.

6.2 Tests Of Significance

A **test of significance** is a procedure for determining the validity of a claim using observed data. The claim is stated in terms of a hypothesis and the result of the statistical test is given in terms of a probability. The hypotheses come in pairs, a null hypothesis or H_0 and an alternative hypothesis or H_a. The probability is stated as a p value and the smaller the p value, the stronger our evidence against H_0. This section describes how to examine various properties of tests of significance.

Test Statistics

Throughout this section of IPS, the authors use the National Loan Survey to illustrate the calculations. The same calculations will be illustrated here using the data from Exercise 6.57 for Degree of Reading Power (DRP). These data can be retrieved from the text Web site.

For this example, our null hypothesis is that there is no difference between the mean of these scores and the national mean (given as 32). The alternative hypothesis (a two-sided alternative) is that there is a difference. These can be formalized as:

   H_0: there is no difference in the true means
   H_a: the true means are not the same

To test your hypotheses, first obtain the mean and standard deviation for your sample data and check for outliers and normality. The steps are repeated for you below.

1. Click **Analyze ▸ Descriptive Statistics ▸ Explore.**

2. Click on **drp** and then ‣ to move **drp** into the "Dependent List" window.
3. Click "**Plots**" in the lower right corner, and then click on "Normality Plots with Tests."
4. Click **Continue** and then click **OK.**

To verify that there are no major departures from normality, check the stem-and-leaf for outliers and then look at the Normal Q-Q Plot to see if all of the points are near the line. The SPSS output shows no reason for concern about outliers or normality. See Table 6.2 for a portion of the SPSS output.

			Statistic	Std. Error
DRP	Mean		35.09	1.687
	95% Confidence Interval for Mean	Lower Bound	31.69	
		Upper Bound	38.49	

Table 6.2

SPSS has given us the mean for the DRP scores (35.09) and the standard error (1.687).

To complete our test of the hypotheses, it is necessary now to calculate a z statistic. The z statistic tells us how far $\bar{x}$ is from μ in standard error units.

To calculate the z statistic and its associated p value, follow these steps:

1. Enter the sample *mean, mu,* and *SE* in the SPSS for Windows Data Editor.
2. Using the **Transform** ‣ **Compute** features, calculate $z = (\textbf{\textit{mean}} - \mu)/\textbf{\textit{SE}}$ (see Figure 6.9).
3. Click **OK** and SPSS will return the value 1.83.

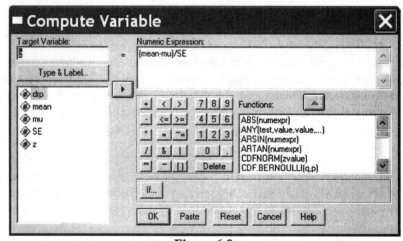

Figure 6.9

p Values

If all test statistics were normal, we could base our conclusions on the value of the z test statistic. Because not all test statistics are z's (random variables with the $N(0;1)$ distribution), we translate the value of test statistics into a common language, the language of probability.

To calculate the probability of getting a z value less than 3.23, use the **CDF.NORMAL** function in SPSS. To get this value, use the following steps:

1. Click on **Transform** ▸ **Compute** and the "Compute Variable" window will appear.
2. Label your target variable (*p,* for example).
3. Choose the **CDF.NORMAL** function by locating it in the "Functions:" window and then double-clicking it to move it into the "Numeric Expression:" window.
4. Replace the first **?** with the *value for z* or the *variable name* for the location of *z* in your SPSS for Windows Data Editor. Replace the second **?** with the mean of a *z* distribution (*0*) and replace the third **?** with the value of the standard deviation for the *z* distribution (*1*) (see Figure 6.10).
5. Click **OK.** SPSS will return the value .96650, which can be rounded to .97.

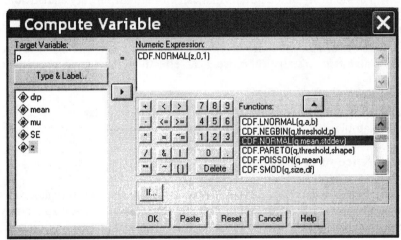

Figure 6.10

Recall that CDF means a Cumulative Distribution Function and therefore gives the probability of every *z* value less than 1.83 for this example. In this example, what you want is the probability of *z* equal to or greater than 1.83. Therefore, to complete our calculations we must do the following:

$$p = p(z > 1.83) = 1 - .97 = .03.$$

To complete our statistical analysis, we must set our significance level or α. In general practice, α is set at .05 (or 1−.95 from our 95% confidence interval). If the *p* value is as small as or smaller than α we say that the data are **statistically significant at level** α. This probability .03 obtained for our DRP example gives us good evidence *against* H_0 (because it is less than .05) and in favor of H_a.

Two-Sided Significance Tests And Confidence Intervals

To look at confidence intervals let us continue with our DRP scores from Exercise 6.57. Refer to Table 6.1 earlier in this chapter. This is only a portion of the output that SPSS generates when we ask for exploratory (**Explore**) data analysis. The standard error of the mean is shown as 1.687. To compute the lower bound of the 95% confidence interval, we need to go down (subtract) 2 standard errors and to compute the upper bound, we go up (add) 2 standard errors. SPSS has computed these values for us and included them in the table. When you do the calculations yourself, your answer will be slightly different than the one shown here. This is due to the fact that SPSS carries 16 decimal places in all calculations but has only given us three decimal places for the standard deviation in the output.

6.3 Use And Abuse Of Tests

Completing a significance test using statistical software such as SPSS is very simple today. However, it remains the responsibility of the user to make decisions along the way. It is essential to declare your hypotheses as a first step. Keep in mind that for some questions that you might ask, a lack of significance is as important as in other situations where you might be looking for significance. Careful statements of your hypotheses are important. After stating your hypotheses, carefully consider your alpha level and how you will use it before you do your exploratory data analysis and inferential statistics. By following a clear protocol, you will avoid many of the pitfalls that can ensnare the unwary statistician. It is also important at the exploratory data analysis stage to confirm that your data meet the assumptions of the statistical test that you plan to run. Choose an alternative test if the assumptions on which the test is based are not met. Consider using confidence levels either in place of or together with your test of significance. Keep in mind that no statistical process can effectively overcome a poorly designed data collection process.

Above all, remember that there is often a very big difference between a statistically significant outcome and a practically meaningful one.

6.4 Power And Inference As A Decision*

Power represents your ability to reject a false null hypothesis in favor of the alternative. Power is of interest to scientists analyzing the data that result from their experiments because generally they are interested in rejecting the null hypothesis. To illustrate the concept of power, we will use Example 6.28 from IPS which looks at total body bone mineral count (TBBMC) for young women.

Power

The following steps illustrate the power calculations for the TBBMC example. The calculations for power are done in three steps.

At **Step 1** we are asked to state H_0, H_a and the significance level.

At **Step 2**, we are asked to write the rule for rejecting H_0 in terms of the $\bar{x}$. The researchers state for us that $\sigma = 2$. We can use SPSS to generate the minimum value of $\bar{x}$ required to reject H_0 at the $\alpha = .05$ level as follows:

1. Enter the *Critz*, *sigma*, and *n* in the SPSS for Windows Data Editor.
2. Click **Transform, Compute** and in the "Compute Variable" window, label your "Target Variable" (e.g. *Xbar*), then enter the calculation equation in the "Numeric Expression:" window (see Figure 6.11).
3. Click **OK.**

SPSS returns the value .66 (rounded to 2 decimal places by default).

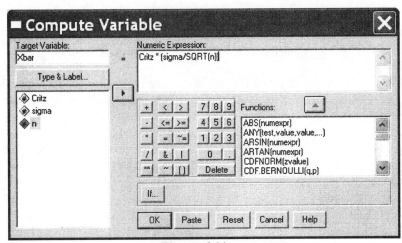

Figure 6.11

In **Step 3**, we calculate power. Power is the probability of this event under the condition that the alternative $\mu = 1$ is true. To calculate this we need to standardize $\bar{x}$ using $\mu = 1$. To calculate power, follow these steps:

1. Click **Transform, Compute** and in the Compute Variable window, label your Target Variable (e.g. *Minz*), then enter the calculation equation in the "Numeric Expression" window (see Figure 6.12). Click **OK.** SPSS returns the value .85, that is, the minimum *z* value required to reject the null hypothesis at the .05 level.

Next ask SPSS to generate the probability of a *z* value of at least .85.

1. Click **Transform, Compute** and in the "Compute Variable" window, label your Target Variable in the "Target Variable" box, as *prob.*
2. In the "Functions" box, click ▼ until **CDFNORM(?)** appears in the "Functions" box. Double-click on **CDFNORM(?)** to move **CDFNORM(?)** into the "Numeric Expression:" box. The **CDFNORM(?)** function stands for the cumulative distribution function for the *z* distribution, and it calculates the area *to the left* of *z* under the *z* distribution. Replace the (?) with the calculated value for *Minz.* Click **OK.**

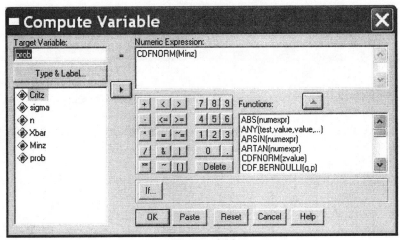

Figure 6.12

SPSS returns the value .20. Because SPSS gives us the p value to the left of the minimum z value, we subtract this from 1 ($1-.20=.80$) to arrive at our power level.

As noted in IPS, it is increasingly common for scientists to be required to include the value of power (80% is the standard) along with the 95% confidence interval and 5% α level when reporting statistical outcomes.

Before closing the SPSS spreadsheet that you have generated for this exercise, change the values of α, sample size, and σ in a stepwise fashion to change the power for this example. See IPS for how to increase power, as this is the most desirable outcome.

Exercises For Chapter 6

1. For each hypothesis below, state whether it is the null or alternative hypothesis and whether it is a one-sided or two-sided hypothesis.
   a. Men are more religious than women.
   b. There is a difference in sensitivity between mothers and fathers.
   c. College students are not more arrogant than the adult population.
   d. Blue collar workers are less alienated than white collar workers.
   e. There is no difference in political apathy between working women and nonworking women.

2. Shown below is a hypothetical set of scores for calcium intake in mg. The data include scores for 30 individuals. Answer the following questions about the data.
   a. Do a complete exploratory data analysis including a Q-Q plot.
   b. Report any outliers, and the mean, standard deviation for calcium intake.
   c. We are told that sigma for calcium intake is 1500. Calculate the margin of error for these data. Explain how doubling your sample size might change (or not) your outcome.
   d. What is the 99% confidence interval for the population mean? For the sample mean?
   e. Write the appropriate two-sided hypotheses to test the difference between the sample mean and the population mean.
   f. Calculate the appropriate inferential statistic and report the value of the statistic and the p value that goes with it.
   g. Do you reject or fail to reject the null hypothesis?

1377.89	1096.79	96.09
334.40	937.28	1265.34
806.16	1099.70	1385.29
592.94	980.33	1100.90
791.62	790.17	511.57
439.36	1279.07	733.57
1416.18	954.22	1928.42
1504.72	951.41	638.40
403.86	1057.47	718.69
911.62	416.40	1378.87

3. Shown on the next page is a hypothetical set of LSAT scores. The data include scores for 40 individuals. Answer the following questions about the data.
   a. Do a complete exploratory data analysis including a Q-Q plot.
   b. Report any outliers, and the mean, standard deviation for LSAT scores.
   c. The cut score (pass/fail score) for this writing of the test was 242. How many "failures" do we have?
   d. We are told that sigma for the LSAT is 100. Calculate the margin of error for these data. Explain how changing z* might change (or not) your outcome.
   e. What is the 90% confidence interval for the population mean of LSAT scores? For the sample mean?
   f. Write the appropriate two-sided hypotheses to test the difference between the sample mean and the population mean.
   g. Calculate the appropriate inferential statistic and report the value of the statistic and the p value that goes with it.
   h. Do you reject or fail to reject the null hypothesis?

597	456	417	668
389	546	413	508
676	492	296	574
548	360	452	367
434	587	304	396
489	452	328	397
614	410	537	477
604	401	571	526
207	576	444	374
671	459	538	674

4. A population is known to have a mean of 41.01 and standard deviation of 12.12. Suppose that three samples of size 36 are drawn from the population. The sample means are 38.55, 39.79, and 37.48.
    a. For each sample find the 90% confidence interval for the population mean.
    b. Does each interval estimate include the known population mean? Is this what you would expect? Why?
    c. Repeat parts a. and b. for a 95% confidence interval.
    d. Calculate power for the following circumstances:
        1. when $n = 36$ and $\alpha = .10$
        2. when $n = 200$ and $\alpha = .20$
        3. when $n = 50$ and $\alpha = .01$
        4. when $n = 10$ and $\alpha = .01$

5. Suppose it is known that, for a particular model car, gasoline consumption is normally distributed with a mean of 43 MPG and a standard deviation of 3 MPG. Four test sites randomly select 16 cars and observe mileage performance in MPG. The data are shown in the table below for each test site.

	N	Minimum	Maximum	Mean	Std. Deviation
Site A	16	39.21	47.10	43.0422	2.92381
Site B	16	32.30	45.74	41.6217	3.38096
Site C	16	36.15	49.61	41.4833	3.25724
Site D	16	36.29	50.74	43.5543	3.53195

    a. What is the probability that an observation will be greater than 52 mpg?
    b. What is the probability that the sample mean at any of the four test sites will be greater than 44 mpg?
    c. Construct an 80% confidence interval for the population mean for each of the four test sites. Briefly describe what this tells you.
    d. Calculate power for the following circumstances:
        1. when $n = 16$ and $\alpha = .05$
        2. when $n = 2$ and $\alpha = .01$
        3. when $n = 100$ and $\alpha = .05$
        4. when $n = 50$ and $\alpha = .20$

6. A large steel plant employs 6828 production employees. Suppose a random sample of 42 workers reveals an average hourly rate of $18.30. Assume that it is known that $\sigma = 3.00$.
    a. Find a 90% confidence interval for the average wage rate for production workers in this steel plant.
    b. A particular individual earns $15.85 per hour. What would you tell her about how her wage compares to that of fellow employees? To complete this problem, calculate the z statistic and its related p value.
    c. What is the probability of earning $29.00 per hour or more?

d. What percentage of employees earns less than $10.00 per hour?
e. Calculate the sample size needed to be 95% confident that our confidence interval contains the actual population mean.
f. Calculate power for the following circumstances:
   1. when n = 42 and α = .01
   2. when n = 5 and α = .20
   3. when n = 100 and α = .05
   4. when n = 42 and α = .05

7. A randomly selected sample of 15 students is asked to report their grade point averages (GPA). Assume that σ = 3.0. The reported data are shown in the table below.
   a. Do a complete exploratory data analysis.
   b. Find the 95% confidence interval for the population mean.
   c. A particular individual reports a GPA = 2.34. What would you tell her about how her GPA compares to that of fellow students? To complete this problem, calculate the z statistic and its related p value.
   d. What is the probability of earning a GPA of less than 1.5?
   e. What percentage of students earns GPA's of at least 2.8?
   f. Calculate the sample size needed to be 85% confident that our confidence interval contains the actual population mean for GPA.
   g. Calculate power for the following circumstances:
      1. when n = 15 and α = .05
      2. when n = 5 and α = .05
      3. when n = 100 and α = .20
      4. when n = 1000 and α = .05

3.24	1.98	3.98
2.76	2.42	2.65
3.02	2.87	3.98
2.28	2.34	3.62
3.62	2.82	3.98

Chapter 7. Inference For Distributions

Topics covered in this chapter:

7.1 Inference For The Mean Of A Population

This section introduces the use of the **t distribution** in inferential statistics for the mean of a population. When σ is known for the population we use the z statistic as learned in Chapter 6. When σ is unknown for the population, the t distribution, rather than the z distribution, is used.

The One-Sample t Confidence Interval

Example 7.1 in IPS explores the Vitamin C content for a sample of corn soy blend (CSB). Compute a 95% confidence interval for μ where μ is the mean Vitamin C content in the population of CSB produced. Before proceeding with the confidence interval for the mean, we must verify the assumption of normally distributed data and compute the $\bar{x}$ and s for the data set.

The SPSS for Windows Data Editor contains a single variable called *Vit_C,* which is declared type numeric 8.0.

To obtain a confidence interval for μ, follow these steps:

1. Click **Analyze,** click **Descriptive Statistics,** and then click **Explore** and the "Explore" window appears.
2. Click *Vit_C,* then click ▸ to move *Vit_C* into the "Dependent List" box.
3. By default, a 95% confidence interval for μ will be computed. To change the confidence level, click **Statistics**. The "Explore: Statistics" window opens and here you can change 95 in the "Confidence Interval for Mean" box to the desired confidence level.
4. Click **Continue.**
5. Click "Plots" in the lower right corner and then click on "Normal Probability Plots with Tests" to obtain the Normal Q-Q plot.
6. By default, the "Display" box in the lower left corner of the "Explore" window has "Both" selected. Click **Statistics** and click the options on or off to meet your requirements.

7. Click **OK.**

Table 7.1 contains the resulting SPSS for Windows output. We are 95% confident that the mean Vitamin C content lies somewhere between 16.4880 and 28.5120 mg/100 g (obtained from the "Lower Bound" and "Upper Bound" rows in Table 7.1).

Descriptives

			Statistic	Std. Error
Vit_C	Mean		22.5000	2.54250
	95% Confidence Interval for Mean	Lower Bound	16.4880	
		Upper Bound	28.5120	
	5% Trimmed Mean		22.6667	
	Median		22.5000	
	Variance		51.714	
	Std. Deviation		7.19126	
	Minimum		11.00	
	Maximum		31.00	
	Range		20.00	
	Interquartile Range		13.75	
	Skewness		-.443	.752
	Kurtosis		-.631	1.481

Table 7.1

The Stem-and-Leaf plot for these data is shown in Figure 7.1. Note that the values are the reverse of those shown in IPS for this example. Otherwise, SPSS has computed exactly the same values as those shown in the text.

```
           Vitamin C Stem-and-Leaf Plot


       Frequency       Stem & Leaf

          2.00          1 .  14
          4.00          2 .  2236
          2.00          3 .  11

       Stem width:      10.00
       Each leaf: 1 case(s)
```
Figure 7.1

The One-Sample *t* Test

To illustrate one-sample *t* procedures, we will continue with the CSB example. According to Example 7.2 in IPS, we know that the intent of the manufacture is to have 40 mg/110g of Vitamin C. We can test the hypothesis that our sample meets this objective. That is, we test the null hypothesis H_0: $\mu = 40$ against H_a: $\mu \neq 40$. Alternatively, we could write a one-directional alternative hypothesis as H_a: $\mu < 40$ or H_a: $\mu > 40$. The two-sided alternative hypothesis is generally recommended as being more conservative (reducing the likelihood of an incorrect decision about whether or not to reject the null hypothesis).

Before proceeding with the one-sample *t* test, it is wise, even for this small sample, to verify the assumption of normally distributed data. To obtain a Q-Q plot and/or stemplot for the variable of interest, follow the directions given above and in earlier chapters of this manual. According to a stemplot, shown above, there is some indication that the distribution takes a normal shape. Further, we are confident that random samples taken continuously in this manner will take a normal distribution.

To conduct a one-sample *t* test, follow these steps:

1. Click **Analyze,** click **Compare Means,** and then click **One-Sample T Test.** The "One-Sample T Test" window in Figure 7.2 appears.

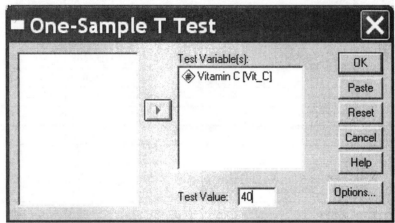

Figure 7.2

2. Click *Vit_C*, then click ▸ to move *Vit_C* to the "Test Variable(s):" box.
3. Type the value of μ_0 (the value of μ under H_0) into the "Test Value:" box. For this example, 40 is the correct value.
4. By default, a 95% confidence interval for $\mu - \mu_0$ will be part of the one-sample *t* test output. To change the confidence level, click **Options,** and then change 95 in the "Confidence Interval" box to the desired confidence level; click **Continue.**
5. Click **OK.**

Table 7.2 is part of the resulting SPSS for Windows output.

One-Sample Test

	Test Value = 40					
					95% Confidence Interval of the Difference	
	t	df	Sig. (2-tailed)	Mean Difference	Lower	Upper
Vit_C	-6.883	7	.000	-17.50000	-23.5120	-11.4880

Table 7.2

Notice that the *t* value is exactly as calculated in IPS. We are testing a two-sided hypothesis, therefore, the significance level shown in Table 7.2 is appropriate for our decision-making. Based on this outcome, we reject H_0 in favor of H_a. From the descriptives shown in Table 7.1, we can see that the mean of this sample has a Vitamin C content of 22.5 mg/110g, on average. Our statistics tell us that this is significantly different from the production specifications.
Example 7.4 in IPS is another illustration. The output, using the protocols mentioned above, is shown below in Figures 7.3 and 7.4 as well as Table 7.3. Compare these to the outcomes shown in IPS for this example.

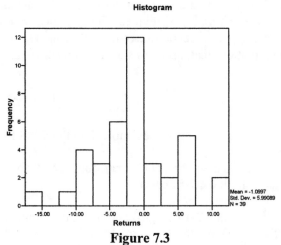

Figure 7.3

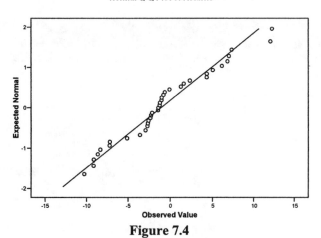

Figure 7.4

One-Sample Test

	Test Value = .95					
					95% Confidence Interval of the Difference	
	t	df	Sig. (2-tailed)	Mean Difference	Lower	Upper
Returns	-2.137	38	.039	-2.04974	-3.9918	-.1077

Table 7.3

Table 7.4 on the next page shows a portion of the Descriptives output that gives us the 95% confidence interval for the investor's returns.

Descriptives

			Statistic	Std. Error
Returns	Mean		-1.0997	.95931
	95% Confidence Interval for Mean	Lower Bound	-3.0418	
		Upper Bound	.8423	

Table 7.4

The following example shows how to generate the exact *P*-value for the one-sample *t* test when summarized data rather than raw data have been provided. For example, we might be interested in the one-sample *t* statistic for testing H_0: $\mu = 10$ versus one of the following, H_a: $\mu < 10$, H_a: $\mu > 10$, or H_a: $\mu \neq$ 10 from a sample of $n = 23$ observations that has a test statistic value of $t = 2.78$. Using software, find the exact *P*-value.

To obtain the *P*-value, follow these steps:

1. Enter the value of the test statistic (2.78 for this example) into the SPSS for Windows Data Editor under the variable called *teststat*.
2. From the SPSS for Windows main menu bar, click **Transform** and then click **Compute.** The "Compute Variable" window in Figure 7.5 appears.
3. In the "Target Variable" box, type in *pvalue.*
4. In the "Functions" box, click ▼ until **CDF.T(q,df)** appears in the "Functions" box. Double-click on **CDF.T(q,df)** to move **CDF.T(?,?)** into the "Numeric Expression:" box. The **CDF.T(q,df)** function stands for the cumulative distribution function for the *t* distribution, and it calculates the area *to the left* of *q* under the correct *t* distribution.

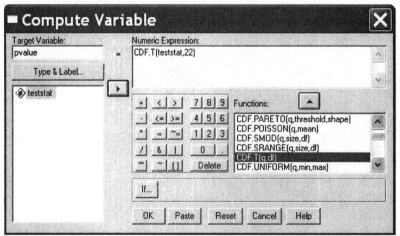

Figure 7.5

5. In the "Numeric Expression:" box, change the first **?** to *teststat* and the second **?** to **22** (n − 1). For our H_a: $\mu < 10$, **CDF.T(*teststat*,22)** should appear in the "Numeric Expression:" box. For H_a: $\mu > 10$, **1 − CDF.T(*teststat*,22)** should appear in the "Numeric Expression:" box. For H_a: $\mu \neq 10$, **2*(1 − CDF.T(ABS(*teststat*),22))** should appear in the "Numeric Expression:" box.
6. Click **OK.**

The *P*-value can be found in the SPSS for Windows Data Editor. By default, the number of decimal places for the variable *pvalue* is two. The number of digits after the decimal place can be changed in the variable view window. The *P*-value for H_a: $\mu < 10$ is 0.995. The *P*-value for H_a: $\mu > 10$ is 0.005. The *P*-value for H_a: $\mu < 10 = .99$.

Matched Pairs *t* Procedures

It is sometimes the case that we have two measurements that can be matched for an individual. Example 7.7 in IPS is this type of example. The data are for three days surrounding full moons and the other days of the month. There are 15 dementia patients in the study. The average number of disruptive behaviors per day was calculated for "Moon days" and "Other days." The difference between the number of disruptive behaviors on the two types of days are included in Table 7.2 of IPS. Instructions are given below as if these differences had not been calculated for us in advance.

To create the variable *diff,* follow these steps:

1. Click **Transform,** and then click **Compute.** The "Compute Variable" window, shown below in Figure 7.6, appears.

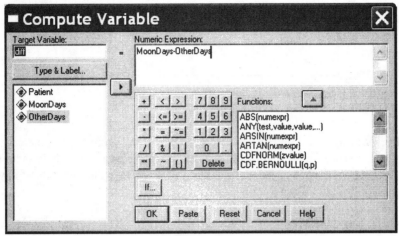

Figure 7.6

2. In the "Target Variable" box, type *diff.*
3. Double-click *MoonDays*, click the gray minus sign (–), and then double-click *OtherDays*. The expression *MoonDays - OtherDays* appears in the "Numeric Expression:" box.
4. Click **OK.** The variable *diff* appears in the SPSS for Windows Data Editor.

Obtain a Q-Q plot and/or stemplot for the variable *diff.* A stemplot of the differences shows that the distribution is reasonably symmetric and appears reasonably normal in shape. The normal Q-Q plot, shown on the next page in Figure 7.7, however, shows three individuals with very small differences in the number of disruptive behaviors. This poses a potential problem for us and we should proceed with caution.

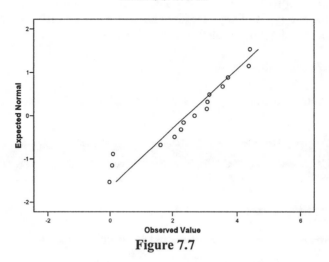

Figure 7.7

To analyze the difference scores, set up H_0: $\mu = 0$ versus H_a: $\mu \neq 0$, where μ = the mean difference in the population from which the subjects were drawn. The null hypothesis says that there is no difference in the number of disruptive behaviors, and the alternative hypothesis says that the type of day does make a difference in the number of disruptive behaviors.

To conduct a matched pairs *t* test, follow these steps:

1. Click **Analyze,** click **Compare Means,** and then click **Paired-Samples T Test.** The "Paired-Samples T Test" window appears.
2. Click ***MoonDays*** to have the expression ***MoonDays*** appear opposite Variable 1 in the "Current Selection" box and then ***OtherDays*** to have ***OtherDays*** appear opposite Variable 2. See Figure 7.8.

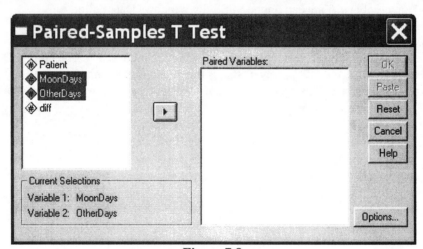

Figure 7.8

3. Click ▶ to have the ***MoonDays - OtherDays*** appear in the "Paired Variables" box.
4. By default, a 95% confidence interval for μ will be part of the matched pairs *t* test output. To change the confidence level, click **Options,** and then change 95 in the "Confidence Interval" box to the desired confidence level.
5. Click **Continue.** Click **OK.**

Table 7.5 is part of the resulting SPSS for Windows output.

Paired Samples Test

		Paired Differences							
					95% Confidence Interval of the Difference				
		Mean	Std. Deviation	Std. Error Mean	Lower	Upper	t	df	Sig. (2-tailed)
Pair 1	MoonDays - OtherDays	2.43267	1.46032	.37705	1.62397	3.24137	6.452	14	.000

Table 7.5

Note: The same results would have been obtained if we had applied the one-sample *t* test to the variable *diff* using a test value of 0.

These data support the claim that dementia patients exhibit more aggressive behaviors in the days around a full moon. In addition, we are 95% confident that the mean difference in disruptive behaviors in the population from which the subjects were drawn lies somewhere between 1.62397 and 3.24137.

To complete the calculations following Example 7.8 in IPS, first use SPSS to give you the critical *t* value as follows:

1. From the SPSS for Windows main menu bar, click **Transform** and then click **Compute.** The "Compute Variable" window appears.
2. In the "Target Variable" box, type in *CritT.*
3. In the "Functions" box, click ▼ until **IDF.T(p,df)** appears in the "Functions" box. Double-click on **IDF.T(p,df)** to move **IDF.T(?,?)** into the "Numeric Expression:" box. The **IDF.T(p,df)** function stands for the inverse cumulative distribution function for the *t* distribution, and it returns the critical value under the correct *t* distribution.
4. In order to calculate the confidence interval, we want the one-sided critical *t* value. Therefore substitute .05/2 = .025 for the p and 14 for the df so that the calculation looks like this: **IDF.T(.025,14).** SPSS returns the value 2.14 (ignore the negative sign in front of the *t* value) or 2.145 as shown in the example in IPS. You can then complete the calculations as shown or return to Table 7.5, above, for the lower and upper confidence limits.

The Sign Test For Matched Pairs

As an alternative to transformations, we can proceed with a nonparametric test, that is, one that does not assume a normal distribution. The Sign Test is one such alternative. To calculate a Sign Test for the data in Example 7.7 for Moon Days versus Other Days, follow these steps:

1. Click **Analyze, Nonparametric, 2 Related Samples** as shown in Figure 7.9 on the next page.
2. Click *MoonDays* and *OtherDays* to move them into the "Test Pair(s) List:" box.
3. Remove the ✔ beside "Wilcoxon" and place it beside "Sign." See Figure 7.10.
4. Click **OK.**

Figure 7.9

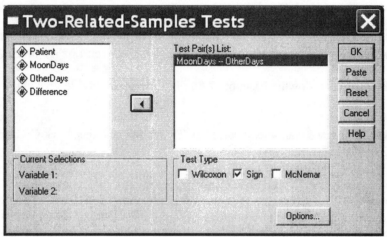

Figure 7.10

The output from SPSS is shown below in Table 7.6. The *p* value given by SPSS does not match exactly that in the text, however, we will come to the same conclusion based on the SPSS output and that is to reject the null hypothesis.

Test Statistics(b)

	OtherDays - MoonDays
Exact Sig. (2-tailed)	.001(a)

a Binomial distribution used.
b Sign Test

Table 7.6

The Power Of The *t* Test

Power represents your ability to reject a false null hypothesis in favor of the alternative. It is based on your sample size and effect size. The effect size is based on the proportion of the total variance that is described by the independent variable(s). Power is of interest to scientists analyzing the data that result from their experiments because generally they are interested in rejecting the null hypothesis. Often the number of subjects chosen is of critical importance. To illustrate the concept of power, we will return to the disruptive behaviors of dementia patients.

Example 7.9 in IPS illustrates the power calculations for H_o: $\mu = 0$ versus H_a : $\mu \neq 0$.

First, following the instructions given above, have SPSS generate the critical t value when p is .95 and df = 19. SPSS returns the value 1.7291 or 1.729 as in the text. We then proceed to compute power. To obtain the P-value, follow the steps described earlier in this chapter. The calculations for power are done in two steps.

For Step 1, we are asked to write the rule for rejecting H_0 in terms of the $\bar{x}$. We know that $\sigma = 1$, so the z test rejects H_0 at the $\alpha = .05$ level when $\bar{x}$ is $\geq$ .520. We can use SPSS to generate this value of .520 as follows:

1. Click **Transform, Compute,** and in the "Compute Variable" window, label your "Target Variable" (e.g. *sampmean*), then enter the calculation equation as shown in Example 15.5 in the "Numeric Expression:" window (see Figure 7.11).
2. Click **OK.**

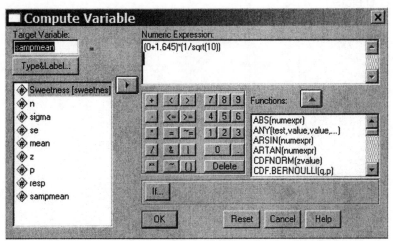

Figure 7.11

In Step 2, we calculate power. Power is the probability of this event under the condition that the alternative $\mu = 1.1$ is true. To calculate this we need to standardize $\bar{x}$ using $\mu = 1.1$. To calculate power, follow these steps:

1. Click **Transform, Compute,** and in the Compute Variable window, label your Target Variable (e.g. *prob*), then enter the calculation equation (as shown in Example 7.9 in IPS) in the "Numeric Expression" window.
2. Click **OK.**

Inference For Nonnormal Populations

Often we can "fix" a skewed variable by transforming the data using logarithms. Example 7.10 in IPS shows us the effects of such a transformation. First we draw a Q-Q Plot using SPSS for the CO emissions. See Figure 7.12 on the next page.

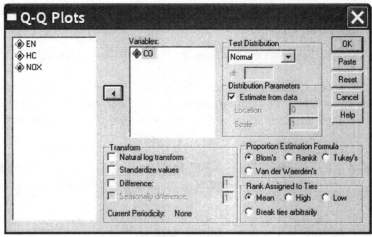

Figure 7.12

Figures 7.13 and 7.15 below and Figure 7.9 in IPS show us that there are a number of nonnormal aspects of these CO emissions data. Recall that if the data are normally distributed, they will fall along the straight line in the Q-Q Plot.

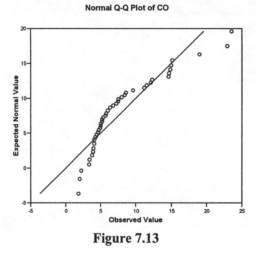

Figure 7.13

The SPSS function **LG10(numexpr)** returns the base-10 logarithm of **numexpr**, which must be numeric and greater than 0.

1. Click **Transform, Compute,** and in the "Compute Variable" window, label your Target Variable (e.g. *CO_Log*).
2. Scroll down through the functions until **Log10(numexpr)** appears and double-click it to move it into the "Numeric Expression" box.
3. Double-click *CO* so that it moves into the equation in place of "numexpr." See Figure 7.14.
4. Click **OK.**

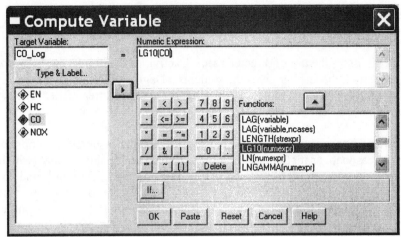

Figure 7.14

The normal Q-Q Plot of ***CO_Log*** is shown in Figure 7.15. The data now more closely resemble a normal distribution and we can proceed with a one-sample *t* test.

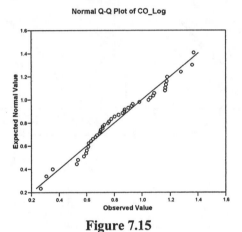

Figure 7.15

7.2 Comparing Two Means

This section of the chapter introduces the use of the *z* and *t* distributions in inferential statistics for comparing two means when the data are for independent samples.

The Two-Sample *z* Statistic

The two-sample *z* statistic is illustrated in Example 7.13 in IPS. It is rare to know σ for both of our populations, therefore, exact *z* procedures are rarely used and SPSS does not provide us with this option. To complete the calculation illustrated in Example 7.13 refer to the text and use the **Compute** function in SPSS or use your calculator.

The Two-Sample *t* Procedures

The two-sample problem examined in this section compares the responses in the two groups, where the responses in each group are independent of those in the other group. Assuming the two samples come from normal populations, the **two-sample *t* procedure** is the correct test to apply. It is also of interest to ask if the variances differ from one group to the next, and this can be tested using the **F test for equality of variances** (see section 7.3 later in this manual for a more detailed discussion).

Example 7.14 in IPS discusses the use of directed reading activities in a third-grade classroom. At the end of an eight-week period, the teacher gives the students a Degree of Reading Power (DRP) test. Did the directed reading activities improve scores for the students given this treatment? The data are contained in Table 7.4 in your text and can be obtained from the text Web site.

Before proceeding with the two-sample t test, we must verify that the assumption of normally distributed data in both groups is reasonably satisfied. Both populations are reasonably normal (which can be determined using boxplots and/or Q-Q plots for the two groups of data). Q-Q Plots and boxplots are shown below in Figure 7.16. Notice that, in the Q-Q plots, there appear to be two low outliers in the Treatment Group and one high outlier in the Control Group. These are not shown as outliers in the boxplots. Thus we can assume normality and continue with our analyses.

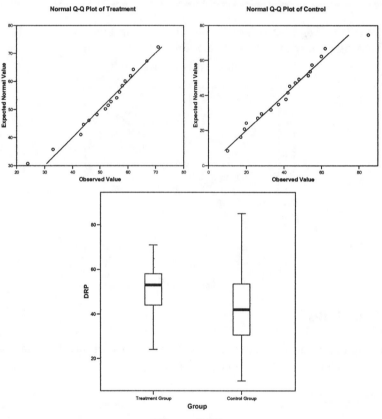

Figure 7.16

Set up H_0: $\mu_1 = \mu_2$ versus H_a: $\mu_1 > \mu_2$, where μ_1 = the Treatment Group and μ_2 = the Control Group. To perform the two-sample t test, follow these steps:

1. If the DRP scores are not stacked in a single column with the group designation in a second column, arrange your data this way first.
2. Click **Analyze, Compare Means,** and then **Independent-Samples T Test.** The "Independent-Samples T Test" window in Figure 7.17 appears.

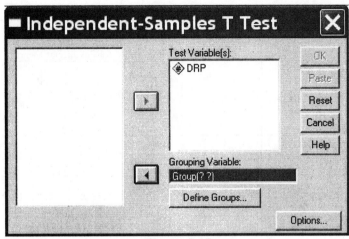

Figure 7.17

3. Click *DRP,* then click ▸ to move *DRP* into the "Test Variable(s)" box.
4. Click *Group,* then click ▸ to move *Group* into the "Grouping Variable" box.
5. Click **Define Groups.** The "Define Groups" window in Figure 7.18 appears.

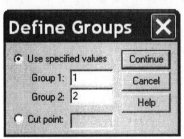

Figure 7.18

6. Type *1* in the "Group 1" box. Press the **Tab** key. Type *2* in the "Group 2" box. *Note:* The group values must be typed in exactly as they appear in the SPSS for Windows Data Editor.
7. Click **Continue. Group (1 2)** appears in the "Grouping Variable" box. By default, a 95% confidence interval for $\mu_1 - \mu_2$ (the difference in population means) will be part of the two-sample *t* test output. To change the confidence level, click **Options** and change 95 in the "Confidence Interval" box to the desired confidence level.
8. Click **Continue.** Click **OK.**

The means and standard deviations for our groups, shown in Table 7.7, match those shown in IPS for Example 7.14. The computed *t* value of 2.311 (Equal variances not assumed; see Table 7.8 on the next page) also matches. To change the probability for the 2-tailed test equal to that of a 1-tailed test, divide by 2. Thus .026/2 = .013, as shown in the text.

Group Statistics

	Group	N	Mean	Std. Deviation	Std. Error Mean
DRP	Treatment Group	21	51.48	11.007	2.402
	Control Group	23	41.52	17.149	3.576

Table 7.7

Independent Samples Test

		Levene's Test for Equality of Variances		t-test for Equality of Means					95% Confidence Interval of the Difference	
		F	Sig.	t	df	Sig. (2-tailed)	Mean Difference	Std. Error Difference	Lower	Upper
DRP	Equal variances assumed	2.362	.132	2.267	42	.029	9.954	4.392	1.091	18.818
	Equal variances not assumed			2.311	37.855	.026	9.954	4.308	1.233	18.676

Table 7.8

A plot of the means is shown in Figure 7.19 below and clearly demonstrates the higher reading scores for the Treatment Group. In Table 7.7 on the next page, we see that the mean difference between our groups is about 10 points (9.954) and this also is demonstrated in the plot of the means. The calculations shown on the next page, however, will tell us that the confidence interval for this difference is rather large, and our data do not permit us to be very precise in our estimate of the size of the improvement on average.

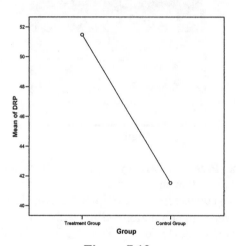

Figure 7.19

The Two-Sample *t* Confidence Interval

Look again at Table 7.8 above. Follow across the line for "Equal variances not assumed" to find the 95% confidence interval for the mean difference between our groups. As in IPS, these values are 1.233 and 18.676. Once again, as noted above, due to the wide confidence interval for the difference between the means, we cannot be very exact in our estimate of the size of the average improvement.

Note: There will often be a slight discrepancy in the confidence interval bounds between IPS and SPSS for Windows. The critical value t^* in the examples requiring hand computations is generally based on a whole number (4, for example), whereas the critical value t^* in SPSS for Windows may be based on exact calculations of df (4.651, for example).

The Pooled Two-Sample *t* Procedures

Example 7.19 and 7.20 discuss the pooled two-sample *t* procedures using Table 7.5 from IPS. Compute the *t* value for two independent groups as shown above for the DRP example. When defining the groups use the appropriate labels as shown in Figure 7.20.

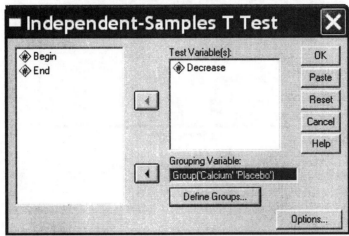

Figure 7.20

The SPSS output is shown below in Tables 7.9 and 7.10. Follow across the row labeled "Equal variances assumed" to get the results shown in IPS. To obtain the *p* value for the one-sided test, divide .119/2 to get .0595 that falls between .05 and .10 as shown in the text.

Group Statistics

	Group	N	Mean	Std. Deviation	Std. Error Mean
Decrease	Calcium	10	5.00	8.743	2.765
	Placebo	11	-.27	5.901	1.779

Table 7.9

Independent Samples Test

		Levene's Test for Equality of Variances		t-test for Equality of Means							
		F	Sig.	t	df	Sig. (2-tailed)	Mean Difference	Std. Error Difference	95% Confidence Interval of the Difference	
									Lower	Upper
Decrease	Equal variances assumed	4.351	.051	1.634	19	.119	5.273	3.227	-1.481	12.026
	Equal variances not assumed			1.604	15.591	.129	5.273	3.288	-1.712	12.257

Table 7.10

To complete the calculations, have SPSS calculate the 90% confidence interval. Follow the instructions given earlier in the chapter except that in the "Options" change the confidence interval to .90 as shown below in Figure 7.21. The SPSS output is shown in Table 7.11 and gives us the same confidence limits as calculated in the text ($-.306$, 10.852).

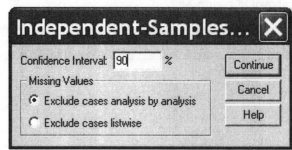

Figure 7.21

Independent Samples Test

		Levene's Test for Equality of Variances		t-test for Equality of Means						
		F	Sig.	t	df	Sig. (2-tailed)	Mean Difference	Std. Error Difference	90% Confidence Interval of the Difference	
									Lower	Upper
Decrease	Equal variances assumed	4.351	.051	1.634	19	.119	5.273	3.227	-.307	10.852
	Equal variances not assumed			1.604	15.591	.129	5.273	3.288	-.477	11.022

Table 7.11

7.3 Optional Topics in Comparing Distributions

The *F* Test for Equality of Spread

SPSS for Windows reports the results of two *t* procedures: the pooled two-sample *t* procedure (assumes equal population variances) and a general two-sample *t* procedure (does not assume equal population variances). To determine which *t* procedure to use, SPSS for Windows performs Levene's Test for Equality of Variances for H_0: $\sigma_1^2 = \sigma_2^2$ versus H_a: $\sigma_1^2 \neq \sigma_2^2$. In Tables 7.10 and 7.11 (on the next page and above), we see an *F* value of 4.351 and a *p* value of .051 under the column headed "Levene's Test for Equality of Variances." Based on the outcome, we fail to reject H_0 in favor of H_a; that is, there is not sufficient evidence at the .05 level to conclude that the population variances are unequal. Thus, we can choose to use the two-sample *t* procedure for which equal variances are assumed. This is an example of a case in statistics where a finding in favor of the null hypothesis is the desired outcome.

Exercises For Chapter 7

1. Based on the following sample:
    a. Do a complete exploratory data analysis.
    b. Estimate the .80 and .99 confidence intervals for the population mean.

8.18
9.93
9.45
9.28
6.29
6.47
9.37
13.27
9.63

2. Based on the following sample:
    a. Do a complete exploratory data analysis.
    b. Estimate the .95 and .99 confidence intervals for the population mean.

160.72	178.68
160.40	174.92
149.34	172.88
184.27	165.19
153.23	153.12
158.96	185.48
140.90	161.40
144.42	164.80

3. Based on the following sample:
    a. Do a complete exploratory data analysis.
    b. Estimate the .95 and .99 confidence intervals for the population mean.

45.05	57.32	59.85	69.32	62.35
56.56	48.97	55.81	67.57	47.26
57.33	54.52	46.69	52.27	56.32
49.92	54.57	47.43	49.82	59.09
62.16	52.75	70.52	59.11	44.40
55.03	63.71	42.52	52.21	50.01
47.28	72.47	58.24	42.71	54.84
65.11	49.20	51.62	54.44	48.84
57.05	47.15	55.59	49.60	43.88
60.69	52.01	65.25	64.74	56.73
57.30	62.08	48.65	62.57	66.93
63.32	58.29	58.47	62.54	59.25
46.62	37.87	61.60	56.65	49.23
45.11	59.91	69.52	53.10	58.54
63.92	58.54	49.43	64.23	66.27
55.25	47.26	58.02	42.28	58.44
52.08	47.92	61.06	60.12	53.42
54.81	63.14	61.32	64.35	64.17
52.03	42.14	69.93	51.84	58.76

4. A new technique has been developed for treating depression. A staff member at a psychiatric hospital tries the technique on a random sample of 9 depressed patients. The population mean length of hospital stay for patients hospitalized for depression is 17 days. The data for length of hospital stay for the patients receiving the new treatment is shown below.
    a. List the hypotheses.
    b. Use $\alpha = .05$.
    c. Do a complete exploratory data analysis.
    d. Is there a difference between the mean length of hospital stay for this treatment group and the population mean length of hospital stay?

1
1
4.94
4.11
6.42
12.30
17.55
20.30
6.88

5. A researcher wants to determine whether the sample mean of a sample of 120 test scores differs from a population mean on the test of 22. The sample data are shown in the table below.
    a. List the hypotheses.
    b. Use $\alpha = .05$.
    c. Do a complete exploratory data analysis.
    d. Is there a difference between the mean score for this treatment group and the population mean?

32.16	27.07	18.81	31.80	25.02	22.85
20.03	23.89	27.55	25.78	22.59	25.48
21.47	15.00	26.02	28.47	29.37	32.35
23.02	25.09	25.68	24.29	23.10	27.40
33.90	25.93	27.73	25.62	32.58	22.51
18.58	24.61	22.23	10.05	25.66	22.90
18.43	27.68	20.66	18.00	20.51	21.04
35.20	24.34	25.67	22.14	26.22	19.66
17.66	18.35	17.03	25.78	17.24	16.20
16.77	12.56	19.06	26.89	18.99	29.93
17.16	19.76	23.48	25.83	23.09	29.91
25.89	20.40	26.58	27.12	32.99	26.15
25.01	26.18	24.56	27.52	16.62	22.41
21.09	30.22	23.27	32.73	26.75	33.73
36.04	27.40	18.60	23.07	25.91	25.24
31.51	20.12	23.43	22.17	26.81	37.72
26.19	20.43	28.38	24.04	19.37	27.84
23.70	25.71	21.13	23.64	21.96	21.64
17.20	24.65	22.66	27.17	15.19	16.58
17.85	26.97	21.74	25.03	19.05	20.64

6. A drug company uses a .05 significance level to reject any lot of medication that does not contain 100 units of a certain chemical. Ten ampoules are selected to test the quality of each of six lots. The resulting mean number of units of the chemical are shown below.
    a. List the hypotheses.
    b. Use $\alpha = .05$.
    c. Do a complete exploratory data analysis.
    d. Which lots should be rejected as having failed to meet quality standards?

Lot 1	Lot 2	Lot 3	Lot 4	Lot 5	Lot 6
98.05	102.77	98.87	100.32	100.55	98.90
97.62	100.08	100.49	100.46	100.02	98.80
102.11	105.02	100.67	100.47	100.65	99.96
101.50	106.39	100.65	100.79	98.98	100.47
101.60	99.70	100.46	100.60	101.36	99.54
97.68	103.54	98.66	100.53	99.69	101.08
103.30	100.49	100.57	100.46	100.85	97.67
100.41	101.76	101.58	100.42	100.47	99.42
99.05	98.32	100.57	100.48	100.87	98.28
104.97	98.19	100.71	99.98	102.77	100.19

7. A team of researchers has developed a test that attempts to predict the level of physical punishment, neglect, and abuse parents will use against their children. A series of observations on the parents with their baby in the newborn nursery, a questionnaire, and interviews are used to determine a "proneness to violence score" (PTV). When the children of these parents are six years old, additional observations, questionnaires, and interviews are conducted and a score determined that reflects the frequency and extent of violence used by the parent against the child. This score is called "violence" (V). The data shown below are for a group of 10 mothers and an independent group of 10 fathers (not the spouses of the mothers sampled). Using $\alpha = .05$, answer the questions shown below. Note that some of the questions require you to return to earlier chapters in the manual to answer. In formulating the answer to each question, list the hypotheses, do a complete exploratory data analysis, and write a summary of the outcome.

    a. Test the hypothesis that there is no relationship between the individual PTV scores and the V scores for mothers. Do the same for fathers.
    b. Test the hypothesis that there is no difference between the mothers and fathers sampled on their PTV scores. Calculate power for this test.
    c. Test the hypothesis that there is no difference between the mothers and fathers sampled on their V scores. Calculate power for this test.
    d. Suppose that the data came from mothers and fathers who were spouses (Mother 1 is the spouse of Father 1). Test the hypothesis that mothers and fathers do not differ in their PTV scores. Did you use the pooled variance test or not? Explain
    e. Suppose that the data came from mothers and fathers who were spouses (Mother a is the wife of Father a). Test the hypothesis that mothers and fathers do not differ in their V scores. Did you use the pooled variance test or not? Explain.
    f. Twice you have tested the hypothesis that mothers and fathers do not differ with respect to PTV scores. Did you arrive at the same conclusion each time? Explain what advantage a matched pairs design might have over an independent groups design.

Mother	PTV	V	Father	PTV	V
1	3	6	1	3	11
2	6	9	2	5	11
3	7	6	3	7	12
4	7	9	4	6	12
5	5	7	5	5	13
6	7	8	6	5	13
7	1	4	7	1	7
8	2	3	8	2	9
9	5	5	9	4	10
10	9	11	10	7	10

Chapter 8. Inference For Proportions

Topics covered in this chapter:

Often we have data that are counts rather than measurements. Chapter 8 in IPS focuses on such data and introduces the concepts related to inference about proportions. The principles and computational techniques are closely related to those introduced in Chapter 7 of the text and manual.

8.1 Inference For A Single Proportion

Large-Sample Confidence Interval For A Single Proportion

Example 8.1 in IPS gives us figures for binge drinking in college students. There are several different ways to approach this problem using SPSS. For this example, use the following instructions:

1. First enter any number in the first 17096 cells of a blank spreadsheet. Use the copy and paste commands to expedite this process.
2. Then generate a bernoulli SRS using the SPSS **Transform** and **Compute** commands as shown in Figure 8.1.

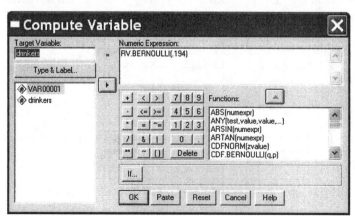

Figure 8.1

3. To obtain the 95% confidence interval, select **Graphs**, **Error Bar**, **Simple**, **Define**. See Figure 8.2.

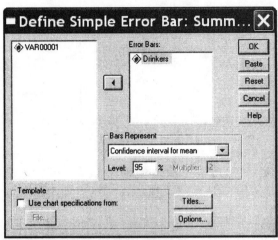

Figure 8.2.

4. Click on "Bars Represent" then choose "Standard error of the mean" and leave the multiplier set at "2". See Figure 8.3.

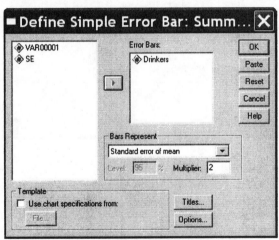

Figure 8.3

5. Click **OK.** The output is shown on the next page in Figure 8.4.
6. Now, following the formulas presented in the text and using the **Transform Compute** options in SPSS, calculate the standard error. See Figure 8.5. The value .03024 is returned.
7. To obtain the confidence interval, follow the instructions in the text; use $z* = 1.960$ and the **Transform Compute** commands in SPSS. The calculations give you a 95% confidence interval of .188 to .200. Look again at Figure 8.4 and confirm these calculations "by eye" from the graph.

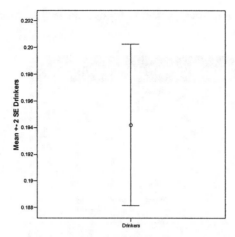

Figure 8.4

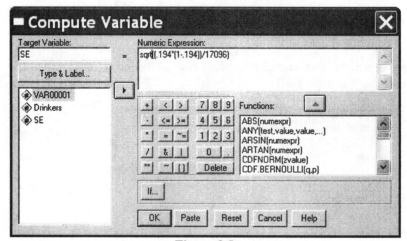

Figure 8.5

Plus Four Confidence Interval For A Single Proportion

The plus four estimation process was suggested by Wilson in 1927. It is used for moderate-sized samples and reduces the inaccuracies that often accompany the large-sample approach discussed earlier. To complete the calculations using this procedure, we add two to the number of successes and four to the sample size. See the text for the rationale behind the approach.

To illustrate the plus four confidence interval for a single proportion, follow Example 8.2 in IPS.

1. First calculate the estimated proportion of equal producers using the **Transform Compute** commands in SPSS.
2. Then compute the standard error as shown above in Figure 8.5 by substituting the appropriate numbers.
3. Now calculate the margin of error as shown in Figure 8.6.
4. Complete the calculations for the confidence interval by adding and subtracting the margin of error from the expected proportion.

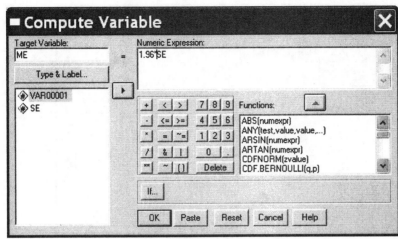

Figure 8.6

Significance Test For A Single Proportion

Example 8.3 in IPS addresses the question: "Does work stress have a negative impact on your personal life?" We can test the responses from a specific workplace where 68% answered "Yes" to the question and the national proportion of 75%. After setting up our hypotheses and assessing our assumptions, as shown in the text, we can calculate a z value. Use the familiar **Transform Compute** commands in SPSS. See Figure 8.7. The value 1.62 is returned as in your text.

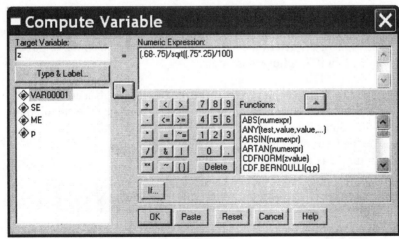

Figure 8.7

After the test statistic is computed, SPSS for Windows can be used to compute the P-value for the test as described in earlier chapters of this manual. Note that the problems in this chapter use the z distribution. As a result, the p values would be computed using the function **1-CDFNORM(?)**. See Figure 8.8. The value returned is .05262.

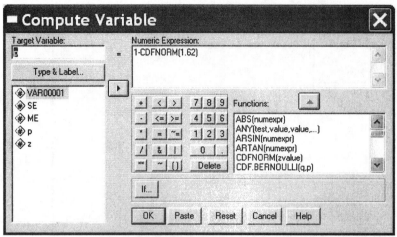

Figure 8.8.

Confidence Intervals Provide Additional Information

Example 8.5 in IPS prompts us to compute a 95% confidence interval for the restaurant worker data. By following instructions given earlier in this chapter of the manual, compute the standard error (see Figure 8.5), margin of error (see Figure 8.6), and 95% confidence interval.

Choosing A Sample Size

The calculation of the appropriate sample size for a proportion is the same as that discussed in Chapter 6. To make the calculations we need p* (often .05), the sample size (n), the population proportion, and the relevant value of $z*$ (determined by the desired confidence interval, often 95%). To replicate Example 8.6 in IPS, follow these steps:

1. Enter the values for m, p, and $z*$ in separate columns of the SPSS spreadsheet.
2. Again we can use the **Transform Compute** commands in SPSS to make the calculations. See Figure 8.9. Note: the double asterisk followed by the number 2 (**2) in the "Expression" window is the instruction to SPSS to square the outcome.
3. Click **OK.** The value returned is 1067.11.

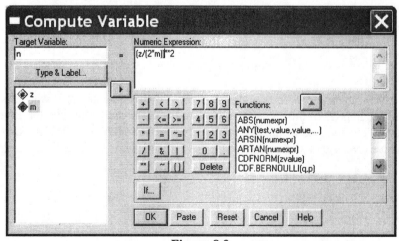

Figure 8.9

We can repeat these figures for a 2% margin of error by following the instructions shown earlier after revising the value of *m* accordingly. For example, to calculate the sample size when the desired margin of error is 4.5%, change *m* to .045 and repeat the calculations. The value returned is 474, or many fewer subjects than when a smaller margin of error is desired.

8.2 Comparing Two Proportions

Large-Sample Confidence Interval For A Difference In Proportions

In Example 8.1 from IPS, we looked at binge drinking in college students. As a follow-up, data were summarized by gender.   We might want to know if the proportion of women in the college population who engage in binge drinking is different from the same proportion for male college students. Once again we will use the **Transform Compute** commands in SPSS to make the calculations.

1. In a blank SPSS spreadsheet enter the values for *n* and *p* for both men and women.
2. Calculate the difference between the two proportions (D) as shown in Figure 8.10.

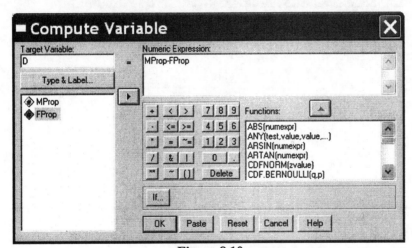

Figure 8.10

3. Calculate the standard error (SE) using the formula shown in the text and in Figure 8.11 following.
4. Calculate *m* following the formula shown in the text.
5. Calculate the upper confidence interval (UCI) by adding *m* to SE as shown in Figure 8.12.
6. Calculate the lower confidence interval by subtracting *m* from SE.

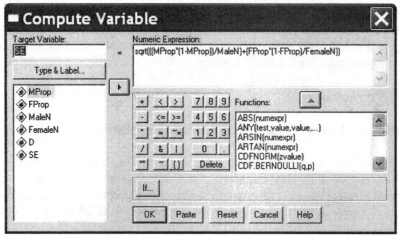

Figure 8.11

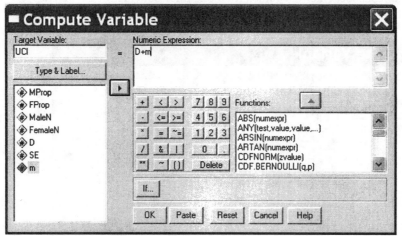

Figure 8.12

Plus Four Confidence Interval For A Difference In Proportions

We can improve the accuracy of our confidence intervals for the difference in proportions by using the Wilson correction as illustrated earlier in the chapter. To do so, we add one success to each sample and add two to the sample size for each sample. To illustrate this approach, the following steps replicate the calculations for Example 8.10 in IPS.

1. Enter the values for Male and Female n and p in a new SPSS spreadsheet.
2. Calculate the plus four estimate of the population proportion for Tanner scores for both boys and girls. The calculation for boys is shown in Figure 8.13.
3. Now calculate the difference (D) between the two proportions (see Figure 8.10, earlier).
4. Calculate the SE of the difference (shown in Figure 8.11 above) following the formula shown in the text.
5. Calculate m and compute the lower and upper limits of the confidence interval as shown earlier in this manual. See Figure 8.12, for example.

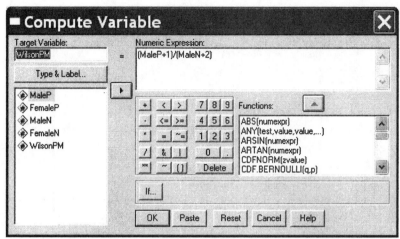

Figure 8.13

Significance Tests

Example 8.11 in IPS addresses the question: "Are men and women college students equally likely to be frequent binge drinkers?" We looked at these data earlier in Example 8.8. Begin by setting up the hypotheses and then, as shown in the text, we can calculate a z value. Use the familiar **Transform Compute** commands in SPSS. See Figures 8.14 through 8.17 for the series of computations using SPSS. The values returned exactly match those in your text.

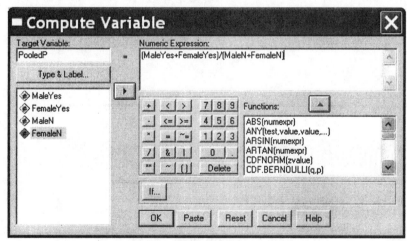

Figure 8.14

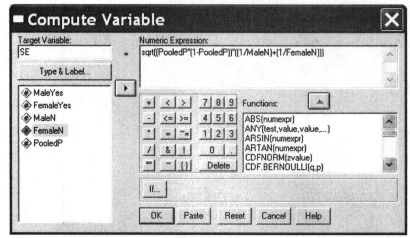

Figure 8.15

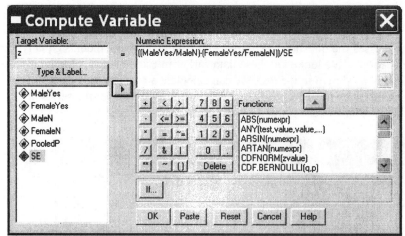

Figure 8.16

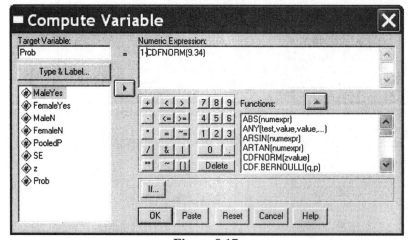

Figure 8.17

Beyond The Basics: Relative Risk

To compare two risks we can calculate the relative risk and then develop the confidence interval for the risk. Example 8.12 in IPS takes us through the computational steps. See Table 8.1 for the organization of the data. Using this organization and notational system will simplify the calculations. Note that the placement of the totals in the appropriate cells is essential to correct calculations using this protocol.

	Binge Drinking	
	Present (Bad Outcome)	**Absent (Good Outcome)**
Men	A = 1630	B = 5550
Women	C = 1684	D = 8232

Table 8.1

Enter these data in four separate columns in SPSS labeled as A through D. Now, begin the process by calculating the relative risk for college men and women of being a frequent binge drinker. See Figure 8.18 for the relative risk (RR) computation where a, b, c, and d refer to cells in the table as described in Table 8.1. The calculation formula is:

$$RR = (A/(A + B)/(C/(C + D)).$$

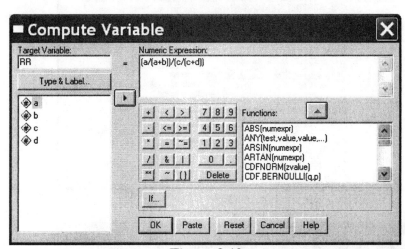

Figure 8.18

We can now go on to compute the lower confidence interval (LCI) for "bad outcomes". By repeating the steps, the upper confidence interval can also be calculated. The following illustration is for the 95% confidence interval. The calculation formula is:

$$RR^{1 \pm (z\alpha/2/\sqrt{X2})}$$

Figures 8.19 through 8.21 illustrate the computations in a stepwise fashion using SPSS for windows beginning with the X^2 calculation, followed by calculation of the lower and upper confidence intervals.

The values returned match those given in IPS Example 8.12. Note that if the confidence interval does not include the value 1.0, the difference in "bad outcome" is considered to be statistically significant for our women and men.

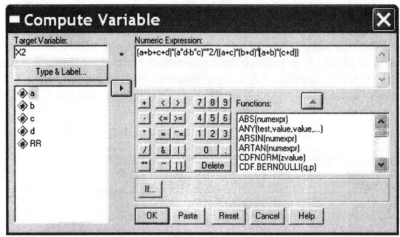

Figure 8.19

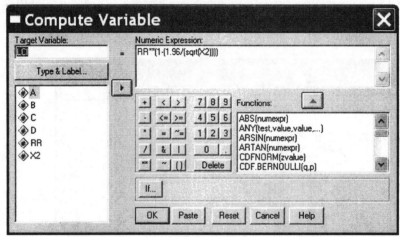

Figure 8.20

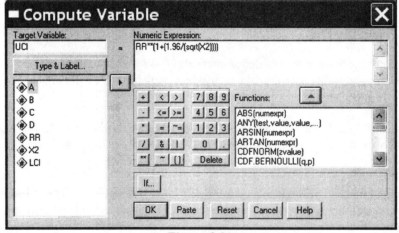

Figure 8.21

Exercises For Chapter 8

1. In a test of a computer aided smoking cessation program, Swiss researchers sent questionnaires to 20,000 residents aged 18 to 60 years. Of the 3124 respondents, 2934 met the criterion of being a daily smoker.
    a. What proportion of the Swiss population could be considered smokers?
    b. Compute the standard error of this proportion.
    c. Calculate the 90% confidence interval for the proportion.

2. A sample of 855 individuals over 30 years of age who reported multiple sexual partners was tested for HIV. Of these, 39 tested positive.
    a. What proportion of this population would you expect to be HIV positive?
    b. Compute the standard error of this proportion.
    c. Calculate the 99% confidence interval for the proportion.

3. Patients who had cardiac surgery performed during the same admission as when they had a symptomatic infection appear to have increased morbidity. Of 13 patients observed under these conditions, 2 died from post-operative complications.
    a. What proportion of this population would you expect to die as a result? Use the Wilson "plus four" technique for moderate sized samples.
    b. Compute the standard error and margin of error for this proportion.
    c. Calculate the 99% confidence interval for the proportion.
    d. How many subjects would be required when the desired margin of error is 2%?

4. Suppose you treated 126 subjects with either a placebo or a new drug designed to decrease the incidence of coronary artery restenosis within 6 months after angioplasty. The number of patients experiencing each outcome is shown in the table below. Is the proportion of placebo-treated versus drug-treated different?
    a. Calculate the difference between the two proportions using the "plus four" method.
    b. Calculate the standard error of the difference.
    c. Calculate the margin of error.
    d. Calculate the upper and lower confidence interval bounds.
    e. Calculate z and the associated p.

		Restenosis within 6 months of surgery?	
		Yes	No
Treatment	Placebo	23	41
	Drug-treated	16	46

5. In a randomized, double-blind clinical trial, oral administration of zidovudine (ZDV), an antiretroviral drug, was initiated at 14-34 weeks gestation in pregnant mothers who tested positive for the AIDS virus. The mother-infant transmission rates were investigated for 364 infants and the outcome is shown in the table on the next page. Is the proportion of placebo-treated versus drug-treated HIV transmission rates different?
    a. Calculate the difference between the two proportions using the "plus four" method.
    b. Calculate the standard error of the difference.
    c. Calculate the margin of error.
    d. Calculate the upper and lower confidence interval bounds.
    e. Calculate z and the associated p.

		Infant HIV?	
		Yes	No
Treatment	ZDV	15	165
	Placebo	47	137

6. Return to the computer aided smoking cessation program described in Question 1. In the table below, you will find the abstinence data (approximated from the research paper). The figures in the table represent self-reported abstinence from tobacco for the previous four weeks, assessed with a survey at seven months after entry into the program. Is the proportion of participants reporting abstinence significantly higher in the intervention group?
    a. Calculate the difference between the two proportions.
    b. Calculate the standard error of the difference.
    c. Calculate the margin of error.
    d. Calculate the upper and lower confidence interval bounds.
    e. Calculate z and the associated p.

		Continuing to use tobacco?	
		Yes	No
Treatment	Intervention	1382	85
	Control	1435	32

7. Beyond the Basics: For questions 4 through 6
    a. Calculate the relative risk.
    b. Calculate the 95% confidence intervals for relative risk.

9.1 Data Analysis For Two-Way Tables

This section introduces the notion of analyzing two or more categorical variables using two-way and multi-way tables. Example 9.1 in IPS takes us back to the binge drinking of college men and women. In the example, we have information for 17,096 individuals. For each individual, we have information about his/her gender (*Gender*), and if he/she is a frequent binge drinker (*BingeDrink*).

These data were entered into an SPSS data spreadsheet using the three variables of *BingeDrink, Gender*, and *Weight* where *Weight* represents the count for a particular combination of *BingeDrink* and *Gender*. Prior to any analyses, the **Weight Cases** option under **Data** was activated and the cases were weighted by *Weight*. See Figure 9.1.

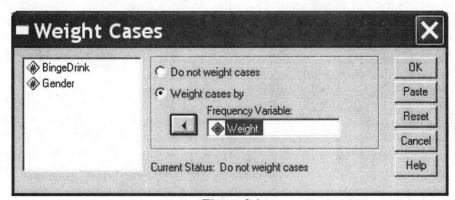

Figure 9.1

To obtain a two-way table similar to that on the page shown with Example 9.2 in IPS, follow the steps shown at the top of the next page:

1. Click **Analyze** ▸ **Descriptive statistics** ▸ **Crosstabs.**
2. Click *BingeDrink,* and then click ▸ to move *BingeDrink* into the "Row(s)" box.
3. Click *Gender,* then click ▸ to move *Gender* into the "Column(s)" box (see Figure 9.2).
4. Click on the "Format" tab in the Crosstabs window and selecting **Descending** under the "Row Order" heading to make the output match the text example.
5. Click **OK.**

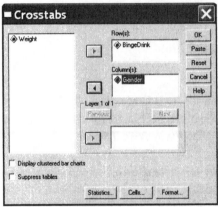

Figure 9.2

The resulting SPSS for Windows output is shown in Table 9.1.

BingeDrink * Gender Crosstabulation

		Gender		Total
		Men	Women	
BingeDrink	No	5550	8232	13782
	Yes	1630	1684	3314
Total		7180	9916	17096

Table 9.1

The choice of row and column variables is important. For the above example we selected "BingeDrink" for the row variable (it could be considered as the dependent variable) and "Gender" for the column variable (considered as if it was an independent variable). In the "Crosstabs" and "Crosstabs: Cell Display" boxes in SPSS, you can change these choices as appropriate. From Table 9.1, we see that 1630/3314 individuals are men who frequently binge drink and 1684/3314 are women who frequently binge drink.

Describing Relations In Two-Way Tables

To **describe the relationship** between two categorical variables, you can generate and compare percentages. For example, we can determine what percents we think would best describe the relationship between gender and frequent binge drinking. Each cell count can be expressed as a percentage of the grand total, the row total, and the column total. To generate these percentages, follow these steps:

1. Click **Analyze** ▸ **Descriptive statistics** ▸ **Crosstabs.**
2. Click *BingeDrink,* and then click ▸ to move *BingeDrink* into the "Row(s)" box.
3. Click *Gender,* then click ▸ to move *Gender* into the "Column(s)" box (see Figure 9.2).
4. Click the **Cells** button. The "Crosstabs: Cell Display" window in Figure 9.3 appears.

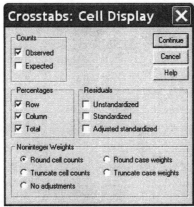

Figure 9.3

5. Click **Row, Column,** and **Total** within the "Percentages" box so a check mark (✔) appears before each type of percentage.
6. Click **Continue** and then click **OK.**

The outcome is shown in Table 9.2.

BingeDrink * Gender Crosstabulation

			Gender		Total
			men	women	
BingeDrink	Yes	Count	1630	1684	3314
		% within BingeDrink	49.2%	50.8%	100.0%
		% within Gender	22.7%	17.0%	19.4%
		% of Total	9.5%	9.9%	19.4%
	No	Count	5550	8232	13782
		% within BingeDrink	40.3%	59.7%	100.0%
		% within Gender	77.3%	83.0%	80.6%
		% of Total	32.5%	48.2%	80.6%
Total		Count	7180	9916	17096
		% within BingeDrink	42.0%	58.0%	100.0%
		% within Gender	100.0%	100.0%	100.0%
		% of Total	42.0%	58.0%	100.0%

Table 9.2

In Table 9.2 above, each cell contains four entries, which are labeled at the beginning of each row. The "Count" is the cell count. The "% within outcome" is the cell count expressed as a percentage of the row total. As an example, out of all 3314 individuals who binge drink, 1630, or 49.2%, are men. The "% within Gender" is the cell count expressed as a percentage of the column total. For instance, out of the 9916 women, 1684, or 23.9%, are frequent binge drinkers. The "% of Total" is the cell count expressed as a percentage of the grand total. Of the 17096 total individuals in the study, 5550, or 32.5%, of the men were not frequent binge drinkers. This output replicates the tables shown in the text for Examples 9.3 to 9.7.

Bar graphs can help us to see relationships between two categorical variables. To make a bar graph comparing the percent of men and women and if they are frequent binge drinkers or not, follow these steps:

1. Click the **Graphs** ▸ **Bar** ▸ **Clustered** ▸ **Define** buttons. The "Define Clustered Bar: Summaries of Cases" window in Figure 9.4 appears.
2. Click "% of Cases" in the "Bars Represent" box.
3. Click *BingeDrink,* then click ▸ to move *BingeDrink* into the "Category Axis" box.
4. Click *Gender,* then click ▸ to move *Gender* into the "Define Clusters by" box.
5. Click **OK.**

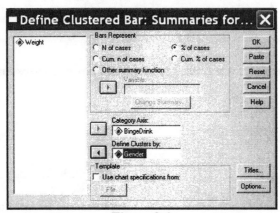

Figure 9.4

The resulting bar graph, shown in Figure 9.5, shows the same outcome as the numerical data; approximately 22% of the frequent binge drinkers are men and about 17% are women.

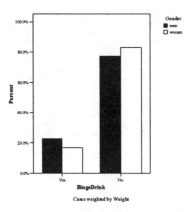

Figure 9.5

Simpson's Paradox

To describe the relationship between three categorical variables, you can generate and compare percentages for different combinations of the data. This will often assist you in analyzing the potential for a lurking variable in the data.   Example 9.10 in IPS is a **three-way table** that reports the frequencies of each combination of levels of three categorical variables: city, airline, and delayed or not. For example, we can determine how many flights are on time from each departure city and add information about which airline was used. To generate these percentages, follow these steps:

1. Activate the **Weight Cases** option under **Data** using *Weight* as the weighting variable.
2. Click **Analyze ▸ Descriptive statistics ▸ Crosstabs.**
3. Click *Airline,* then click ▸ to move *Airline* into the "Row(s)" box.
4. Click *Flight,* then click ▸ to move *Flight* into the "Column(s)" box.
5. Click *City,* then click ▸ to move *City* to the "Layer 1 of 1" box.
6. Click the **Cells** button. The "Crosstabs: Cell Display" window in Figure 9.6 appears.
7. Click **Row, Column,** and **Total** within the "Percentages" box so a check mark (✓) appears before each type of percentage.
8. Click **Continue** and then click **OK.**

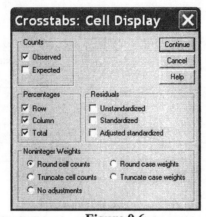

Figure 9.6

The SPSS output is shown on the next page in Figures 9.7 and 9.8 and Table 9.3. Figures 9.7 and 9.8 were generated by first splitting the data file according to city and then forming the graphs. To replicate them, follow these instructions:

1. Click on **Data,** then on **Split File.**
2. Click *City,* then click ▸ to move *City* into the "Groups based on:" box.
3. Click on "Compare Groups".
4. Click **OK.** See Figure 9.9.
5. Click on **Graphs, Bar, Clustered, Define.**
6. Click on "% of cases".
7. Click *Airline,* then click ▸ to move *Airline* into the "Category axis" box.
8. Click *Flight,* then click ▸ to move *Flight* into the "Define clusters by" box. See Figure 9.10.
9. Click **OK.**

Edit the fill color of the bars if you choose to (refer to earlier chapters in this manual if necessary).

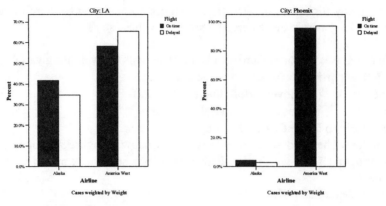

Figure 9.7   Figure 9.8

Airline * Flight * City Crosstabulation

City				Flight		Total
				On time	Delayed	
LA	Airline	Alaska	Count	497	62	559
			% within Airline	88.9%	11.1%	100.0%
			% within Flight	41.7%	34.6%	40.8%
			% of Total	36.3%	4.5%	40.8%
		America West	Count	694	117	811
			% within Airline	85.6%	14.4%	100.0%
			% within Flight	58.3%	65.4%	59.2%
			% of Total	50.7%	8.5%	59.2%
	Total		Count	1191	179	1370
			% within Airline	86.9%	13.1%	100.0%
			% within Flight	100.0%	100.0%	100.0%
			% of Total	86.9%	13.1%	100.0%
Phoenix	Airline	Alaska	Count	221	12	233
			% within Airline	94.8%	5.2%	100.0%
			% within Flight	4.4%	2.8%	4.2%
			% of Total	4.0%	.2%	4.2%
		America West	Count	4840	415	5255
			% within Airline	92.1%	7.9%	100.0%
			% within Flight	95.6%	97.2%	95.8%
			% of Total	88.2%	7.6%	95.8%
	Total		Count	5061	427	5488
			% within Airline	92.2%	7.8%	100.0%
			% within Flight	100.0%	100.0%	100.0%
			% of Total	92.2%	7.8%	100.0%

Table 9.3

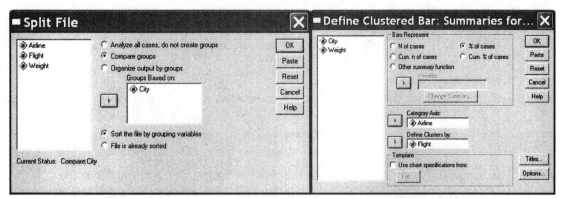

Figure 9.9   Figure 9.10

9.2 Inference For Two-Way Tables

This section introduces significance testing for examining the relationship between two categorical variables. The data often are summarized using a two-way table. **Inferential** procedures are applied to test whether the two categorical variables are independent.

In Example 9.12 in IPS, we are told that many popular businesses are franchises. Contracts are negotiated between the local entrepreneur and the franchise firm. A clause that may or may not be in the contract addresses the entrepreneur's right to an exclusive territory. The objective in this example is to compare franchises that have exclusive territories with those that do not.

The data are displayed in a 2 * 2 contingency table following Example 9.12 in IPS. The subjects were 170 new franchise firms. The entries in this table are the observed, or sample, counts. For example, 108 firms had exclusive territories and were successful. Note that the marginal totals are given with the table; they are not part of the raw data but are calculated by summing over the rows or columns. The row totals are the numbers of observations sampled from the two types of firms and the two types of territories. The grand total, 170, can be obtained by summing the row or the column totals. It is the total number of observations (new franchise firms) in the study.

These data were entered into an SPSS spreadsheet using the three variables of *Success, Territory*, and *Weight* where *Weight* represents the count for a particular combination of *Success* and *Territory*. Prior to any analyses, the **Weight Cases** option under **Data** was activated and the cases were weighted by the variable called *Weight*. To describe the relationship between these two categorical variables, you can generate and compare percentages. Territory is considered the explanatory variable and it is therefore used as the column variable. To generate these percentages, follow these steps:

1. Click **Analyze ▸ Descriptive statistics ▸ Crosstabs.** The "Crosstabs" window in Figure 9.11 appears.
2. Click *Success,* then click ▸ to move *Success* into the "Row(s)" box.
3. Click *Territory,* then click ▸ to move *Territory* into the "Column(s)" box.
4. Click the **Cells** button. The "Crosstabs: Cell Display" window in Figure 9.12 appears.
5. Click beside "Observed", then **Continue,** and then click **OK.**

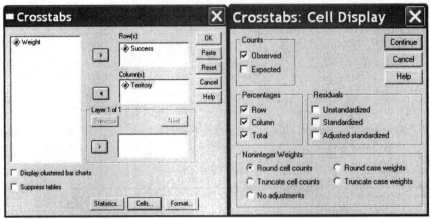

Figure 9.11   **Figure 9.12**

The resulting SPSS for Windows output is shown in Table 9.4.

Success * Territory Crosstabulation

		Territory		Total
		Yes	No	
Success	Yes	108	15	123
	No	34	13	47
Total		142	28	170

Table 9.4

We can add additional information to the outcome by asking SPSS to calculate row and column percentages for us.

1. Repeat Steps 1 through 4 in the previous analysis.
2. Click **Row, Column,** and **Total** within the "Percentages" box so a check mark (✔) appears before each type of percentage. See Figure 9.12 above.
3. Click **OK.** The output is shown in Table 9.5.

Success * Territory Crosstabulation

			Territory		Total
			Yes	No	
Success	Yes	Count	108	15	123
		% within Success	87.8%	12.2%	100.0%
		% within Territory	76.1%	53.6%	72.4%
		% of Total	63.5%	8.8%	72.4%
	No	Count	34	13	47
		% within Success	72.3%	27.7%	100.0%
		% within Territory	23.9%	46.4%	27.6%
		% of Total	20.0%	7.6%	27.6%
Total		Count	142	28	170
		% within Success	83.5%	16.5%	100.0%
		% within Territory	100.0%	100.0%	100.0%
		% of Total	83.5%	16.5%	100.0%

Table 9.5

Each cell contains four entries, which are labeled at the beginning of each row. The "Count" is the cell count. The "% within Success" is the cell count expressed as a percentage of the row total. As an example, out of 123 firms that were successful, 108 (87.8%) had exclusive territories. The "% within Territory" is the cell count expressed as a percentage of the column total. For instance, 108 firms with exclusive territories make up 76.1% of the total number of successful firms. The "% of Total" is the cell count expressed as a percentage of the grand total. Of the 170 total firms who participated in the study, 108, or 63.5%, had exclusive territories and were successful.

The Hypothesis: No Association

The null hypothesis for a two-way table is expressed as: There is no association between the row variable and the column variable; more specifically, the null hypothesis states that success and having an exclusive territory are not related. The alternative hypothesis says that the distributions within the $r * c$ table are not all the same. If there is no association between success and having an exclusive territory, how likely is it that a sample would show differences as large or larger than those displayed in Figure 9.3 of IPS?

Expected Cell Counts

To test the null hypothesis stated above, we compare expected cell counts with those observed. Expected cell counts can be obtained using this simple formula:

$$\text{Expected cell count} = (\text{row total} * \text{column total})/\text{grand total}$$

If calculating these values yourself, be sure that you use the appropriate row and column totals for each cell in the table. Most computer software will complete these calculations for you. To ask SPSS to calculate them for you, in the "Crosstabs: Cell Display" window shown in Figure 9.12 earlier, click "Expected" under "Counts".

The Chi-Square Test

The **chi-square (χ^2) test of independence,** which compares the observed and expected counts, can be used to assess the extent to which the cell distributions can be plausibly attributed to chance. A large value of χ^2 provides evidence against the null hypothesis.

To perform the χ^2 test of independence, follow these steps:

1. Click **Analyze** ▸ **Descriptive statistics** ▸ **Crosstabs.**
2. Click *Success,* then click ▸ to move *Success* into the "Row(s)" box.
3. Click *Territory,* then click ▸ to move *Territory* into the "Column(s)" box. Recall that the explanatory variable is considered the column variable.
4. If you are interested in including the expected cell counts within the contingency table that will result from the analysis, click the **Cells** button and click **Expected** under **Counts** so a check mark (✔) appears before **Expected.** If a check mark appears before **Row, Column,** and **Total** under the "Percentages" box, click on each of these terms so that the check marks disappear.
5. Click **Continue.**
6. Click the **Statistics** button and then click **Chi-Square.**
7. Click **Continue** and then click **OK.**

The resulting SPSS for Windows output is shown in Tables 9.6 and 9.7.

Success * Territory Crosstabulation

			Territory		
			Yes	No	Total
Success	Yes	Count	108	15	123
		Expected Count	102.7	20.3	123.0
	No	Count	34	13	47
		Expected Count	39.3	7.7	47.0
Total		Count	142	28	170
		Expected Count	142.0	28.0	170.0

Table 9.6

Chi-Square Tests

	Value	df	Asymp. Sig. (2-sided)	Exact Sig. (2-sided)	Exact Sig. (1-sided)
Pearson Chi-Square	5.911(b)	1	.015		
Continuity Correction(a)	4.841	1	.028		
Likelihood Ratio	5.465	1	.019		
Fisher's Exact Test				.021	.016
Linear-by-Linear Association	5.876	1	.015		
N of Valid Cases	170				

a Computed only for a 2x2 table
b 0 cells (.0%) have expected count less than 5. The minimum expected count is 7.74.

Table 9.7

Because all the expected cell counts are moderately large, the χ^2 distribution provides accurate p-values. The test statistic value (Pearson Chi-square) is $\chi^2 = 5.91$ with $df = 1$, and $p = 0.015$. The chi-square test confirms that the data contain clear evidence against the null hypothesis that there is no relationship between success and exclusive territories. Under H_0, the chance of obtaining a value of χ^2 greater than or equal to the calculated value of 5.91 is small — less than 15 times in 1000. Note that this result matches the output shown in Figure 9.4 in IPS. You are reminded to ignore the additional output that SPSS provides.

The Chi-Square Test And The z Test

When we have only two populations to compare, we can use either the z test or the χ^2 test. This applies only to 2 * 2 tables since the z test compares only two populations. If we were to apply the techniques learned in Chapter 8 to test the proportions for the current example, we would arrive at the same conclusion. For any 2 * 2 table, the square of the z statistic will equal χ^2.

Beyond the Basics: Meta-analysis

As described in IPS, meta-analysis is a collection of statistical techniques for combining information from different but similar studies. Example 9.17 in IPS gives us the relative risk for eight studies. In each study, young children were given large doses of Vitamin A. The overall results showed that young children in developing countries who were treated with Vitamin A supplements were less likely to die than untreated controls. The relative risk for each of these eight studies is shown in the text. We are asked to compute the relative risk across all eight studies and to compute the 95% confidence interval. To complete this analysis, enter the data in an SPSS spreadsheet and then follow the steps on the next page:

1. Click **Analyze ▸ Descriptive statistics ▸ Descriptives.**
2. Click *RelativeRisk,* then click ▸ to move *RelativeRisk* into the "Variable(s):" box.
3. Click **Options** and click on "Mean" and "S.E. mean".
4. Click **OK.**

SPSS gives us the mean for Relative Risk as 77.00 as shown in the text and the standard error of Relative Risk as .05772. To obtain the 95% confidence interval, use the **Transform, Compute** option to first subtract twice the standard error from the mean and then add the same value to the mean. SPSS returns the values of .6546 and .8854. Note that these values differ slightly from those shown in the text.

9.3 Formulas And Models For Two-Way Tables

Computations

Example 9.18 in IPS takes you through the computations for two-way tables. The steps are reviewed here for SPSS for the same example. First, enter the data in an SPSS spreadsheet or retrieve it from the Web site. In the steps that follow, the variables are called *Wine, Music,* and *Bottles*. The following steps outline the process. First, weight the data by *Bottles* using **Data, Weight Cases,** then use the "Weight cases by" options. Refer to Figure 9.1 earlier if necessary.

Computing Conditional Distributions

1. Calculate the descriptive statistics using **Analyze ▸ Descriptive statistics ▸ Crosstabs.** Move *Wine* into the "Row(s):" box and *Music* into the "Column(s):" box.
2. Click on "Cells:" and then on "Column" in the "Percentages" box.
3. Click on **Continue** and then **OK.**

This is an ideal point to begin a written summary of the relationships seen in the data. Figures 9.6 and 9.7 in IPS show these relationships. To replicate these figures, follow these steps:

1. To begin, click on **Data, Split File,** "Compare Groups" and then move *Music* into the "Groups based on" box. Click **OK.**
2. Click **Graphs, Bar, Simple, Define.**
3. Click on "% of cases" in the "Bars Represent" box.
4. Move *Wine* into the "Category axis" box.
5. Click **OK.**
6. Repeat Steps 1 through 5 moving *Wine* into the "Groups based on:" box and *Music* into the "Category axis" box.

Computing Expected Cell Counts

1. Calculate the expected cell counts using **Analyze ▸ Descriptive statistics ▸ Crosstabs.**
2. Move *Wine* into the "Row(s):" box and *Music* into the "Column(s):" box.
3. Click on "Cells:" and then on "Expected" in the "Counts" box.
4. Click on **Continue** and then **OK.**

To complete the analysis, it is necessary to reverse the split files action completed earlier in the analysis. These steps will complete that action:

1. Click on **Data, Split Files,** and then click on *Music* in the "Groups based on" box and move it back into the variables box.
2. Now click on "Analyze all cases, do not create groups".
3. Click **OK.**

The χ^2 Statistic And Its *p*-value

Finally, we complete the calculations for the χ^2 test and its associated *p* value.

1. Click **Analyze ▸ Descriptive statistics ▸ Crosstabs.**
2. Click on "Suppress tables" then click in the "Statistics" box.
3. Click on "Chi-square".
4. Click **Continue.**
5. Click **OK.**

9.4 Goodness Of Fit

Example 9.24 in IPS gives us an example of the number of automobile collisions by day of the week for 699 drivers who were using a cellular telephone at the time of the accident. The data form a one-way table with seven cells. Note that we cannot use the **Crosstabs** function for this one-way table. To compute the χ^2 goodness of fit test and its associated *p* value for a one-way table, use the following steps:

1. Enter the data in an SPSS spreadsheet or retrieve it from the Web site. Here the variables are called *Day* (a string variable) and *Collisions.*
2. Weight the cases by *Collisions.*
3. Click **Analyze, Nonparametric Tests, Chi-Square.**
4. Move *Collisions* into the "Test Variable List:" box.
5. Click **OK.**

SPSS gives us $\chi^2 = 208.847$, *df* = 6, and *p* = .000. This matches the outcome reported in IPS.

Exercises For Chapter 9

1. The department of psychology offers six separate sections of the introductory level course, Psychology 100. Each section is taught by a different instructor. After final examinations were over, the student association contacted all the students enrolled in the course and asked, "If you had it to do over again, would you take Psychology 100?" The data are shown in the table below.
    a. Write the hypotheses for this question.
    b. Enter the data and then obtain a two-way table of counts and expected counts. Recall that the explanatory variable is entered as the column variable.
    c. Determine the column percentages and write a brief summary of what they tell you about the data.
    d. Make a bar graph describing the data for each section of the course.
    e. Determine the values of $\chi 2$, df, and p.

	Responses to Question	
	Yes	No
Section		
A	16	4
B	16	11
C	19	2
D	14	11
E	31	4
F	18	14

2. The editor of a college newspaper asks a reporter to determine whether student opinion about legalizing abortion is related to the gender of the student. The registrar's office provides the names of 40 male and 40 female students, drawn at random from the registration records. Three students declined to answer the reporter's question about abortion. The survey results are shown in the table below. Answer the following questions about these data.
    a. Write the hypotheses for this question.
    b. Enter the data and then obtain a two-way table of counts and expected counts.
    c. Determine the column percentages and write a brief summary of what they tell you about the data. Recall that the explanatory variable is entered as the column variable.
    d. Make a bar graph describing the data for Males and Females.
    e. Determine the values of $\chi 2$, df, and p.

	Yes	No	Undecided
Males	14	13	12
Females	12	17	9

3. A cognitive psychologist hypothesizes that risk-taking in decision-making is related to a personal characteristic called "gain-loss importance" that is measured by asking people to relate three important events in their lives. If the events are about positive experiences, the person is classified as a person to whom gains are more important than losses. The opposite classification is made if the events are about losses. The psychologist asks 12 students who have been previously classified as gain-important or loss-important to place mock bets on horse races to be run the next day. The students study information about the race in the *Daily Racing Form* and place their bets according to odds given in the *Form*. The amount of risk-taking is determined by the person's tendency to play long-shots. Higher scores indicate more risky choices. According to the following scores, can gain-loss importance predict risk-taking?
    a. Write the hypotheses for this question.
    b. Enter the data and then obtain a two-way table of counts and expected counts.

c. Determine the column percentages and write a brief summary of what they tell you about the data. Recall that the explanatory variable is entered as the column variable.
d. Determine the values of $\chi 2$, df, and p.

	Scores
Gain-important	12, 14, 9, 22, 13, 17, 14
Loss-important	8, 18, 11, 10, 15, 12, 14

4. Theories of adolescence suggest that young people are primarily concerned with establishing their own goals, values, and so on. A researcher considering this idea hypothesizes that if it is true then young people should exhibit apathy about the plight of others. Below are (hypothetical) scores from an apathy measure administered to four age groups. Lower scores indicate high apathy while higher scores indicate a greater concern for others.
   a. Write the hypotheses for this question.
   b. Enter the data and then obtain a two-way table of counts and expected counts.
   c. Determine the column percentages and write a brief summary of what they tell you about the data. Recall that the explanatory variable is entered as the column variable.
   d. Make a bar graph describing the data for each age group.
   e. Determine the values of $\chi 2$, df, and p.

Age Group	Scores
21-30	13, 9, 11, 12, 10, 8
31-40	11, 12, 9, 10, 14, 16
41-50	11, 15, 12, 14, 10, 12
51-60	10, 12, 16, 13, 9, 14

5. Suppose you conducted a drug trial on a group of animals and you hypothesized that the animals receiving the drug would survive better than those that did not receive the drug. You conduct the study and collect the data shown below.
   a. Write the hypotheses for this question.
   b. Enter the data and then obtain a two-way table of counts and expected counts.
   c. Determine the column percentages and write a brief summary of what they tell you about the data. Recall that the explanatory variable is entered as the column variable.
   d. Determine the values of $\chi 2$, df, and p.

	Dead	Alive
Treated	36	14
Not Treated	30	25

6. Return to Question 5 and:
   a. Write the hypotheses for this question.
   b. Compute the proportion of animals that survived in each of the groups.
   c. Compute a z value and its associated p value according to the techniques learned in Chapter 8 of this manual.
   d. Square the z value and compare it to the χ^2 value computed in Question 5.

7. Suppose you performed a simple monohybrid cross between two individuals that were heterozygous for the trait of interest, Aa by Aa. The results of your cross are shown in the table below.
   a. Write the hypotheses for this question.
   b. Enter the data and then obtain a two-way table of counts and expected counts.
   c. Determine the column percentages and write a brief summary of what they tell you about the data. Recall that the explanatory variable is entered as the column variable.

d. Determine the values of χ2, *df,* and *p.*

	A	a
A	10	42
a	33	15

8. Return to Question 7 and:
    a. Write the hypotheses for this question.
    b. Compute the proportion of crosses for each of the combinations.
    c. Compute a *z* value and its associated *p* value according to the techniques learned in Chapter 8 of this manual.
    d. Square the *z* value and compare it to the χ^2 value computed in Question 7.

9. The controversy over the possible association between magnetic field exposure and childhood leukemia has led to several research studies. One summary of these studies is presented in the table below (http://www.powerlinefacts.com/EMF%20Metatstudy.pdf.) In this case, we can treat the odds ratio the same way as we have treated relative risk in this chapter.
    a. Write the hypotheses for this question.
    b. Calculate the mean, standard error, and 95% confidence interval for the odds ratio of any type of leukemia for children exposed to magnetic fields.

Study First Author	Odds Ratio
Tomenius	.34
Myers	1.56
Savitz	1.93
London	1.68
Feychting	2.49
Olsen	1.5
Verkasalo	1.55
Linet	1.19
Tynes	.27
Michaelis	2.74
McBride	1.35
Dockerty	2.71
Green	1.39
UK	.96

10. Four cohort studies have been published on cancer risks in the children of mothers who smoked during pregnancy. These data are summarized in the table below.
     a. Write the hypotheses for this question.
     b. Calculate the mean, standard error, and 95% confidence interval for Relative Risk of any type of cancer for infants whose mother smoked during pregnancy.

Country, period of diagnosis	Study size, years of follow up	Relative Risk
Canada & UK, 1959–1968	89,302, 7–10 years	1.3
UK, 1970–1980	16,193, 0–10 years	2.5
Sweden, 1982–1987	497,051, 0–5 years	1.0
US, 1959–1974	54,795, 0–8 years	.75

11. You have noticed that pine trees grow well in some parts of the woods, but not others. You speculate that the distribution of pines is related to drainage, that is, that pines prefer a very well drained soil, while they do poorly in wet areas. You sample soil from evenly spaced plots throughout the forest, two days after a heavy rain. You discover that there are three *categories* of soil: dry (sample falls

apart in your hand), loamy (holds shape if you squeeze it, falls apart if you drop it), and wet (muddy—you can squeeze lots of water out, soil is muddy). The table below shows the results for 100 plots. Does soil drainage have any effect on the distribution of pines?

a. Write the hypotheses for this question.

b. Enter the data and then obtain a two-way table of counts and expected counts.

c. Determine the values of $\chi 2$, *df,* and *p*.

Soil Type	Plots with Pines
Dry	31
Loamy	17
Wet	2

12. Students at a small university were asked to describe the socio-economic status of the community in which they grew up. The results are shown in the table below.

a. Write the hypotheses for this question.

b. Enter the data and then obtain a two-way table of counts and expected counts.

c. Determine the values of $\chi 2$, *df,* and *p*.

Affluent	11
Upper Middle Class	50
Middle Class	32
Lower Middle Class	8
Working Class	10
Poor	1

Chapter 10. Inference for Regression

Topics covered in this chapter:

The descriptive analysis discussed earlier in Chapter 2 for relationships between two quantitative variables leads to formal inference. This chapter focuses on demonstrating how SPSS for Windows can be used to perform inference for simple linear regression and correlation.

10.1 Simple Linear Regression

For the examples in this chapter, we will use the following data. Infants who cry easily may be more easily stimulated than others and this may be a sign of higher IQ at 3 Years. Child development researchers explored the relationship between the crying of infants 4 to 10 days old and their later IQ test scores. A snap of a rubber band on the sole of the foot caused the infants to cry. The researchers recorded the crying and measured its intensity by the number of peaks in the most active 20 seconds. They later measured the children's IQ at age 3 using the Stanford-Binet IQ test. Table 10.1 contains data on 38 infants. To replicate the examples in this chapter, type these data into an SPSS spread sheet.

Crying	IQ	Crying	IQ	Crying	IQ	Crying	IQ
10	87	20	90	17	94	12	94
12	97	16	100	19	103	12	103
9	103	23	103	13	104	14	106
16	106	27	108	18	109	10	109
18	109	15	112	18	112	23	113
15	114	21	114	16	118	9	119
12	119	12	120	19	120	16	124
20	132	15	133	22	135	31	135
16	136	17	141	30	155	22	157
33	159	13	162				

Table 10.1

Statistical Model For Linear Regression

The response variable is the IQ at 3 Years test score and is plotted on the y axis. The explanatory variable is the count of crying peaks and is plotted on the x axis. Figure 10.1 is the scatterplot of the data set with the fitted line superimposed. There is a moderate positive linear relationship, with no extreme outliers or

161

potentially influential observations. As a result, it is now of interest to model this relationship. Compare this plot to Figures 10.3 and 10.4 in IPS and the considerations made there.

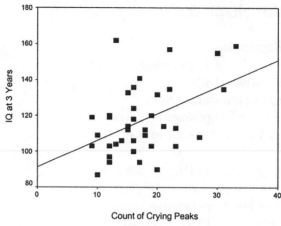

Figure 10.1

To obtain the correlation and the least-squares regression line, follow these steps:

1. Click **Analyze,** click **Regression,** and then click **Linear.** The "Linear Regression" window in Figure 10.2 appears.

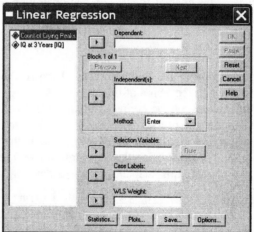

Figure 10.2

2. Click *IQ at 3 Years,* then click ▸ to move *IQ at 3 Years* into the "Independent(s)" box.
3. Click *Count of Crying Peaks,* then click ▸ to move *Count of Crying Peaks* into the "Dependent" box.
4. If you are interested in a normal quantile plot for the residuals, click the **Plots** box (Figure 10.3 on the following page appears). Select the **Normal probability plot** option in the "Standardized Residual Plots" box.
5. Click Continue.
6. If you are interested in saving predicted values, residuals, distance measures, influential statistics, and prediction intervals, click the **Save** box (Figure 10.4 appears). Check the desired options and click **Continue.**

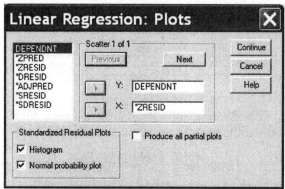

Figure 10.3

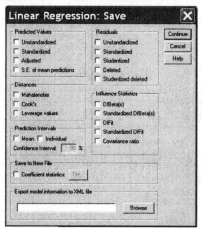

Figure 10.4

7. If you are interested in obtaining the confidence interval for the slope, click the **Statistics** box (Figure 10.5 appears). Check **Confidence intervals** in the "Regression Coefficients" box.
8. Click **Continue** and then click **OK.**

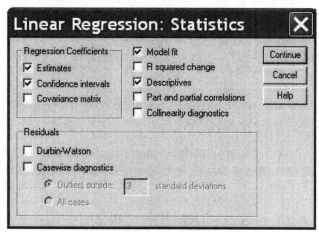

Figure 10.5

Portions of the output from SPSS, beginning with descriptive statistics, are shown in the tables on the next page, beginning with Table 10.2.

Descriptive Statistics

	Mean	Std. Deviation	N
IQ at 3 Years	117.24	19.383	38
Count of Crying Peaks	17.39	5.907	38

Table 10.2

Tables 10.3 and 10.4 are part of the resulting SPSS for Windows output. All the data required to construct the regression formula including the slope and intercept are shown here.

Model Summary(b)

Model	R	R Square	Adjusted R Square	Std. Error of the Estimate
1	.455(a)	.207	.185	17.499

a Predictors: (Constant), Count of Crying Peaks
b Dependent Variable: IQ at 3 Years

Table 10.3

Coefficients(a)

Model		Unstandardized Coefficients		Standardized Coefficients	t	Sig.	95% Confidence Interval for B	
		B	Std. Error	Beta			Lower Bound	Upper Bound
1	(Constant)	91.268	8.934		10.216	.000	73.149	109.388
	Count of Crying Peaks	1.493	.487	.455	3.065	.004	.505	2.481

a Dependent Variable: IQ at 3 Years

Table 10.4

Simple Linear Regression Model

Table 10.3 gives us the correlation between our variables (.455). The correlation of $r = .455$ indicates that a moderately strong, positive, linear association exists between IQ at 3 Years and Counts of crying peaks. In Table 10.4 we can find the y-intercept and the slope in the column labeled "Unstandardized Coefficients"; these values are required to construct our regression equation. The equation for the least squares regression line is:

$$IQ \text{ at } 3 \text{ Years} = 91.268 + 1.493 * Count \text{ of Crying Peaks}$$

See Example 10.4 in IPS for a similar equation. 91.268 is the mean value of **IQ at 3 Years** when **Count of Crying Peaks** is equal to 0.

Estimating The Regression Parameters

Now that a fitted model has been developed, the residuals of the model should be examined. Recall that the unstandardized residuals (and various other values) were saved into the Data Editor in Step 6. Figure 10.6 displays a portion of the Data Editor containing the values that were saved; the variable **RES_1** contains the residuals. The variable **PRE_1** contains the predicted value for each value of **Count of Crying Peaks**.

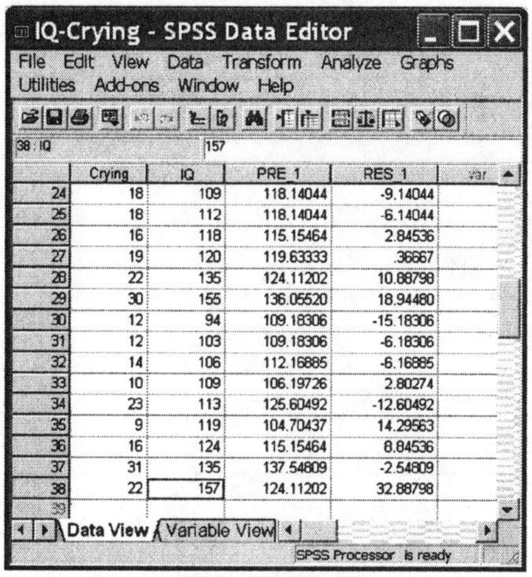

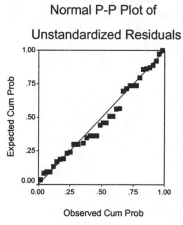

Figure 10.6

Figure 10.7 is a normal quantile plot of the standardized residuals. Because the plotted values are fairly linear, the assumption of normality seems reasonable. For a similar plot see Figure 10.8 in IPS.

Normal P-P Plot of

Unstandardized Residuals

Figure 10.7

In addition, we can examine whether the residuals display any systematic pattern when plotted against other variables. Figure 10.8 is a scatterplot of *RES_1* and ***Count of Crying Peaks.*** No unusual patterns or values are observed. For these data we do not have case number information therefore the relationship between case number and Count of Crying Peaks cannot be examined. It is good practice to record the case number when collecting data.

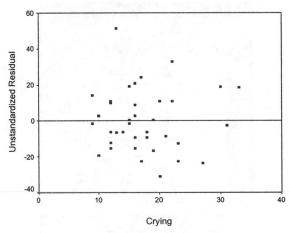

Figure 10.8

Confidence Intervals And Significance Tests

By looking back to Table 10.4 earlier in this chapter, we can find the 95% confidence interval for the intercept and the slope. These values are 73.149 to 109.388 for the intercept and .505 to 2.481 for the slope. The degrees of freedom (df) are 36 and can be found in Table 10.5 below. The test statistic t is computed by SPSS and takes the value of 10.216 (also shown in Table 10.4). This is computed by dividing the intercept (91.268) by the standard error of the intercept (8.934). These values can also be found in Table 10.4 above.

The hypotheses appropriate for testing this relationship are H_0: $\beta = 0$ which says that Count of Crying Peaks has no straight-line relationship with IQ at 3 Years, and H_a: $\beta \neq 0$ which states that there is a straight-line relationship (without specifying positive or negative) between the variables. According to Table 10.4, the computed test statistic $t = 3.065$ and the p-value is .004. Thus there is strong evidence that there is a straight-line relationship between IQ at 3 Years and Count of Crying Peaks.

The same information can be obtained from the SPSS output shown below in Table 10.5. Included here are the degrees of freedom (df) for these data (n-2=36) and an F value. The F value shown is equal to the square of the t value noted above (for your trivia file!).

ANOVA(b)

Model		Sum of Squares	df	Mean Square	F	Sig.
1	Regression	2877.480	1	2877.480	9.397	.004(a)
	Residual	11023.389	36	306.205		
	Total	13900.868	37			

a Predictors: (Constant), Count of Crying Peaks
b Dependent Variable: IQ at 3 Years

Table 10.5

Note that some software allows you to choose between one-sided and two-sided alternatives in tests of significance. Other software always reports the p-value for the two-sided alternative. If your alternative hypothesis is one-sided, you must divide p by 2.

Table 10.4 also shows that a 95% confidence interval for β is (.505, 2.481). See Example 10.6 in IPS for the calculation formulas required to calculate these upper and lower confidence limits.

Confidence Interval For A Mean Response

It is also of interest to construct confidence intervals for the mean response and prediction intervals for a future observation. To have SPSS calculate these for you, set the "Linear Regression: Save" commands as shown in Figure 10.9. Note that both boxes for saving prediction intervals have been checked and that the confidence level is set at the default (95%).

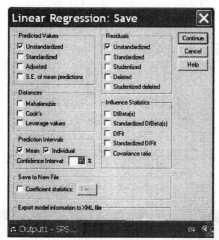

Figure 10.9

Figure 10.10, below, shows the 95% confidence intervals for the mean response for each observation in the data set and 95% prediction intervals for future observations equal to each observation in the data set. The new variables represent the lower (*LMCI_1)* and upper (*UMCI_1)* 95% confidence intervals for the mean response, and the variables (*LICI_1, UICI_1*) represent the lower and upper 95% individual prediction intervals.

IQ-Crying - SPSS Data Editor

File   Edit   View   Data   Transform   Analyze   Graphs   Utilities   Add-ons   Window   Help

38 : IQ   157

	Crying	IQ	PRE_1	RES_1	LMCI_1	UMCI_1	LICI_1	UICI_1
24	18	109	118.14044	-9.14044	112.35240	123.92848	82.18249	154.09839
25	18	112	118.14044	-6.14044	112.35240	123.92848	82.18249	154.09839
26	16	118	115.15464	2.84536	109.23504	121.07425	79.17528	151.13401
27	19	120	119.63333	.36667	113.66192	125.60475	83.64541	155.62126
28	22	135	124.11202	10.88798	116.77491	131.44914	87.87246	160.35159
29	30	155	136.05520	18.94480	122.33853	149.77186	98.00760	174.10279
30	12	94	109.18306	-15.18306	101.33865	117.02747	72.83739	145.52873
31	12	103	109.18306	-6.18306	101.33865	117.02747	72.83739	145.52873
32	14	106	112.16885	-6.16885	105.50656	118.83114	76.05986	148.27784
33	10	109	106.19726	2.80274	96.89740	115.49713	69.50993	142.88460
34	23	113	125.60492	-12.60492	117.61782	133.59203	89.22819	161.98165
35	9	119	104.70437	14.29563	94.61029	114.79845	67.80771	141.60102
36	16	124	115.15464	8.84536	109.23504	121.07425	79.17528	151.13401
37	31	135	137.54809	-2.54809	122.92907	152.16712	99.16595	175.93023
38	22	157	124.11202	32.88798	116.77491	131.44914	87.87246	160.35159
39								
40								

Data View Variable View

SPSS Processor is ready

Figure 10.10

Figure 10.9 in IPS shows the 95% confidence limits for the mean response for the fuel efficiency example. Figure 10.11 on the next page gives a similar picture for our crying versus IQ data. Notice that there is greater variability in IQ for the higher counts of crying peaks. You can replicate this output by following the SPSS instructions for Line Charts as shown in Figure 10.12. Notice that the order in which the variables appear in the "Lines Represent" box is important so that the legend is organized logically.

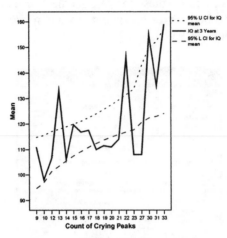

Figure 10.11

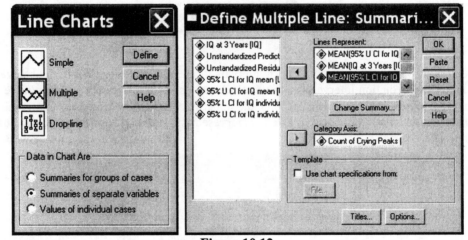

Figure 10.12

Prediction Intervals

We can follow a similar process to illustrate prediction intervals for future observations. Earlier we asked SPSS to save the prediction intervals and the variables (*LICI_1*, *UICI_1*) represent the lower and upper 95% individual prediction intervals (see Figure 10.10, above). The 95% prediction limits for the individual responses are shown in Figure 10.13 (see also Figure 10.10 in IPS).

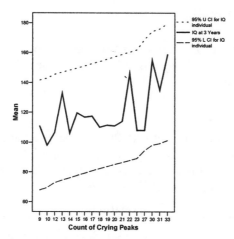

Figure 10.13

Beyond The Basics: Nonlinear Regression

Nonlinear regression is a method of finding a nonlinear model of the relationship between the dependent variable and a set of independent variables. Unlike traditional linear regression, which is restricted to estimating linear models, nonlinear models can estimate models with arbitrary relationships between independent and dependent variables. In order to complete your calculations, it is first necessary to specify the appropriate regression model. There are many published models and a full description is beyond the scope of this text.

Example: Can population be predicted based on time? A scatterplot shows that there seems to be a strong relationship between population and time, but the relationship is nonlinear, so it requires the special estimation methods of the Nonlinear Regression procedure. By setting up an appropriate equation, such as a logistic population growth model, we can get a good estimate of the model, allowing us to make predictions about population for times that were not actually measured.

10.2 More Detail About Simple Linear Regression

Analysis Of Variance For Regression

In Table 10.5 earlier, we saw a typical ANOVA output from SPSS for a regression analysis. Refer to that table in this section. The total variation in our example is equal to 13900.868 and this is often referred to as the sums of squares total or SST. The sums of squares for the model (SSM) is equal to 2877.480 and the sums of squares error or residual (SSE) is 11023.389. Similarly, we can find the associated df in Table 10.5 and df total is 37.

When we were first looking at our regression equation, we looked for the value of the correlation r between our variables. Look back at Table 10.3 and locate the values for r (.455) and r^2 (.207). Following our instruction in IPS we find that SSM/SST will give us r^2 also. Confirm these calculations for yourself.

Mean squares (MS) are calculated by dividing SS's by their associated df. Thus, in our example, MS for the model (MSM) or MS for the regression is equal to 2877.480/1 = 2877.480 and MS error (MSE) is 11023.389/36 = 306.205. When MS for the model is divided by MSE, we obtain an F value; in this case it is 9.397. Large values of F are evidence against our null hypothesis. In this case, based on our p value of .004 (which is less than .05) we conclude that our F value is large enough to reject the null hypothesis that there is no relationship between our variables in favor of an alternative hypothesis that says there is.

Calculations For Regression Inference

In these sections of IPS, the authors take you through a step-by step protocol for completing regression computations using your calculator. You may wish to replicate these calculations using a small subset of the IQ-Crying data used in this chapter.

Inference For Correlation

Returning to our data for crying and IQ, we can ask SPSS to calculate our correlation (ignore the fact that it already appears in Table 10.3 earlier in this chapter). SPSS will print out the correlation together with a t value and significance level. Follow the instructions in Chapter 2 for completing these calculations. A segment of the SPSS printout is shown below in Table 10.6.

Correlations

		Count of Crying Peaks	IQ at 3 Years
Count of Crying Peaks	Pearson Correlation	1	.455(**)
	Sig. (2-tailed)		.004
	N	38	38
IQ at 3 Years	Pearson Correlation	.455(**)	1
	Sig. (2-tailed)	.004	
	N	38	38

** Correlation is significant at the 0.01 level (2-tailed).

Table 10.6

Given that our p level (.004) is less than .05, we can conclude that there is a non-zero correlation between *IQ at 3 Years* and *Count of Crying Peaks*.

Although it is of interest that the correlation between our variables is significant, it is also important to note that the correlation is only moderate and that a review of Figures 10.11 and 10.12 show much variability and predicted variability for future observations. That is, crying at the early age of 4 to 10 days is not a perfect predictor of IQ 3 years later!!!

Exercises For Chapter 10

1. A set of pairs of data is shown below. For these data use SPSS to answer the questions that follow the data.

X	Y
3.68	−.3
7.36	−.6
11.04	−.9
14.72	−1.2
18.4	−1.5
22.08	−1.8
25.76	−2.1
29.44	−2.4
33.12	−2.7
36.80	−3.0

   a. Use the linear regression commands to get descriptive statistics.
   b. Draw and interpret a scatterplot for these pairs of data.
   c. Compute and plot the residuals.
   d. What is the correlation between these variables? Is it significant (use a two-sided alternative)?
   e. What is the value of the slope? The intercept?
   f. Write the regression equation for these variables.
   g. Using the regression equation, predict a value of Y when X = 20; when X = 35.

2. Here is another set of fictional data. Answer the following questions for all possible pairs of numerical variables (e.g. height & weight, height & shoe size, weight & IQ, etc.).

NAME	HEIGHT	WEIGHT	SHOE SIZE	RING SIZE	IQ
George	72	200	12	8	140
Herman	70	160	10	6	130
Bernard	68	170	11	7.5	120
Brent	70	150	8.5	6.5	119
Donald	66	150	8	6	115
Clayton	72	185	11	7.75	114
Graham	66	135	7.5	6	114
Tom	69	165	9	7.5	113

   a. Complete an exploratory data analysis for each variable including both numerical and graphical displays.
   b. Use the linear regression commands to get descriptive statistics.
   c. Draw and interpret a scatterplot for these pairs of data.
   d. Compute and plot the residuals.
   e. What is the correlation between these variables? Is it significant (use a two-sided alternative)?
   f. What is the value of the slope? The intercept?
   g. Write the regression equation for these variables.
   h. Compute confidence intervals and significance tests.
   i. Graph the confidence interval for the mean response for each pair of variables.
   j. Graph the prediction intervals for each pair of variables.
   k. Write the correlation hypotheses (null and alternative) and test them statistically.
   l. Print and interpret the ANOVA summary table for each set of calculations.

m. Write an overall summary of your findings.

3. Review the calculation formulas at the end of Chapter 10 in IPS and then complete the following questions. A researcher tested 2nd grade children for their reading skills. She used a test that had three subscales: decoding, vocabulary, and comprehension. Here's what she found:

	Decoding (D)	Vocabulary (V)	Comprehension (C)
Mean	29.97	11.93	42.87
St. Deviation	7.93	.37	23.19
Correlation	D/V .60	D/C .77	V/C .56

a. Find the value of **b** in the regression equation that predicts:
   i.   Decoding score if you know the student's vocabulary score
   ii.   Vocabulary score if you know the student's comprehension score
   iii. Comprehension score if you know the student's decoding score
   iv.   Decoding score if you know the student's comprehension score

4. Here are some imaginary facts about cockroaches: The average number of roaches per home in the U.S. is 57 with a standard deviation (s) of 12. The average number of packages of roach killer purchased per family per year is 4.2, s = 1.1. These (imaginary) data are from a sample of 1000 homes. The correlation between roaches in the home and roach killer purchased is $r = .5$.
a. The Clean family bought 12 packages of roach killer last year.
   i.   How many roaches would you predict they have in their home?
   ii.   Using your calculation formulas and calculator, find the standard error of the estimate of the number of roaches in their home.
b. The Spans family has 83 roaches in their home.
   i.   Predict how many boxes of roach killer they purchased.
   ii.   Calculate the upper and lower confidence intervals for being 95% correct in your predictions.

5. Here are some fictional facts about language development: The average age for babies born in 1966 said their first sentence at 12.3 months (s = 4.3). Their mothers, born between 1956 and 1976, said their first sentences at age 11.5 months (s = 4.1). These (imaginary) data are from a sample of 1600 mothers and their babies. The correlation between these two variables is $r = .61$.
a. If we know that baby Leslie said her first sentence at 10.8 months, what is our best guess as to when her mother first strung words together in a sentence?
b. Find the standard error of the estimate for the prediction of when a baby will start to talk.
c. Jacob's mother began to talk at 15.8 months. When should she expect Jacob to speak his first sentences?
d. Find the range of ages at which baby Jacob can be expected to start to talk with a 95% chance of being correct.

Chapter 11. Multiple Regression

Topics covered in this chapter:

11.1 Inference For Multiple Regression
11.2 A Case Study
> **Preliminary Multiple Regression Analysis**
> **Relationships Between Pairs Of Variables**
> **Regression Analysis**
> **Interpretation Of Results**
> **Residuals**
> **Refining The Model**

11.1 Inference For Multiple Regression

In this chapter we will demonstrate and discuss some aspects of multiple regression. The data that are used throughout the chapter are available from the IPS Web site associated with Chapter 11. A brief description of the data follows.

The U.S. Federal Trade Commission annually rates varieties of domestic cigarettes according to their tar, nicotine, and carbon monoxide content. The U.S. Surgeon General considers each of these substances hazardous to a smoker's health. Past studies have shown that increases in the tar and nicotine content of a cigarette are accompanied by an increase in the carbon monoxide emitted from the cigarette smoke. The data used here can be accessed from http://www.stat.ucla.edu/data/ and many others can be accessed from similar Web-based data archives.

The following section of the chapter shows the data analysis techniques for multiple regression based on these data. For a theoretical understanding of the analysis, thoroughly review the first section of this chapter in IPS.

11.2 A Case Study

Preliminary Multiple Regression Analysis

To begin the process, conduct the familiar exploratory data analysis for these data. The basic numerical descriptors are shown in Table 11.1 on the next page. Histograms are shown in Figure 11.1 for each variable. These distributions appear to be normal however boxplots give us a clearer visual description of outliers if there are any.

Descriptive Statistics

	N	Minimum	Maximum	Mean	Std. Deviation
Tar (mg)	25	1.00	29.80	12.2160	5.66581
Nicotine content (mg)	25	.13	2.03	.8764	.35406
Weight (g)	25	.79	1.17	.9703	.08772
Carbon monoxide content (mg)	25	1.50	23.50	12.5280	4.73968
Valid N (listwise)	25				

Table 11.1

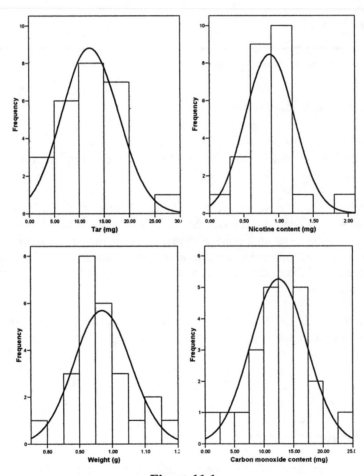

Figure 11.1

Figure 11.2, on the following page, shows boxplots for these variables. From this figure we see that Bull Durham stands out as high on Tar and Nicotine. At the same time, Now cigarettes appear to be lower than most in terms of Nicotine, Weight, and Carbon Monoxide content. These may be considerations later in our analysis.

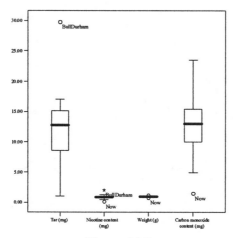

Figure 11.2

Relationships Between Pairs Of Variables

Table 11.2 shows the correlations between all possible pairs of variables. The correlations between Tar and Nicotine, Tar and Carbon Monoxide, and between Nicotine and Carbon Monoxide are very high. Look below at the plots in Figure 11.3 and confirm this interpretation. These high correlations may be an example of collinearity, a factor that may be problematic in our multiple regression analysis but beyond the scope of this discussion.

Variable 1	Variable 2	N	Correlation	P value	Significance
Tar	Nicotine	25	.977	.000	**
Tar	Weight	25	.491	.013	*
Tar	Carbon Monoxide	25	.957	.000	**
Nicotine	Weight	25	.500	.011	*
Nicotine	Carbon Monoxide	25	.926	.000	**
Carbon Monoxide	Weight	25	.464	.019	*

  * Correlation is significant at the 0.05 level (2-tailed).
  ** Correlation is significant at the 0.01 level (2-tailed).

Table 11.2

Scatterplots for each of our bivariate correlations are shown in Figure 11.3 on the next page. The high and the low correlations and the outliers noted above are clearly visible in these plots. It is important to conduct these explorations prior to beginning our multiple regression analysis.

Regression Analysis

When we refer to the story behind these data, presented at the beginning of the chapter, it appears that carbon monoxide is our dependent or response variable and that tar and nicotine, and possibly weight, are variables that explain carbon monoxide content. We are now ready to run the multiple regression analysis. To do so, follow these steps:

1. Click **Analyze,** click **Regression,** and then click **Linear.** The "Linear Regression" window in Figure 11.4 on the next page appears.

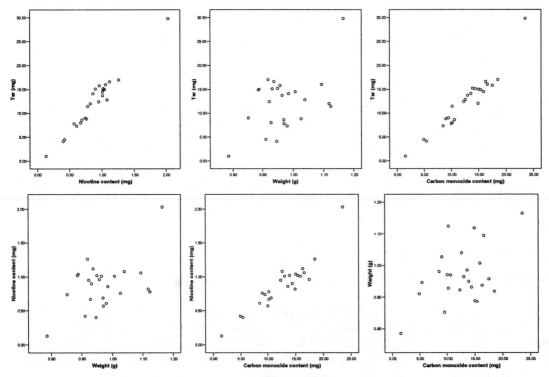

Figure 11.3

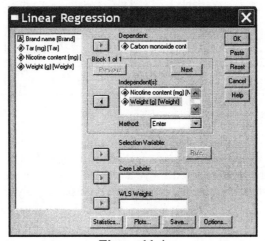

Figure 11.4

2. Move the variables into their relevant boxes: carbon monoxide into the "Dependent:" box and the remaining numerical variables into the "Independent(s):" box.
3. Leave the rest of the settings (default settings) as they are.
4. Click "Statistics" and check the boxes as seen on the next page in Figure 11.5.
5. Click **Continue.**

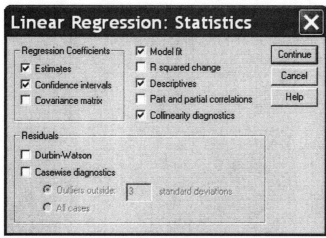

Figure 11.5

6. Click "Plots" and check off any plots that you wish to create. See Figure 11.6. Click **Continue.**
7. Click "Save" and choose the information (residuals in particular) that you wish to have saved into your SPSS spreadsheet. See Figure 11.7 on the next page.
8. Click **Continue.**
9. Click **OK.**

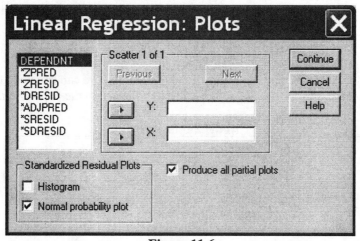

Figure 11.6

Selected portions of the SPSS output are reproduced on the next page. First we have the ANOVA summary table shown in Table 11.3. In the ANOVA table first check the degrees of freedom to ensure that no major errors have been made thus far with data entry or specifying our model. We have 25 cigarette brands in the analysis so degrees of freedom total should be 24 and they are. Since we have 3 explanatory variables, df for Regression equals 3. The residual (24−3 = 21) will always be correct if the other two are.

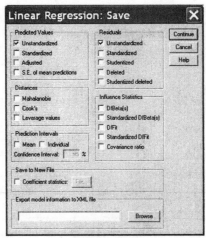

Figure 11.7

ANOVA(b)

Model		Sum of Squares	df	Mean Square	F	Sig.
1	Regression	495.258	3	165.086	78.984	.000(a)
	Residual	43.893	21	2.090		
	Total	539.150	24			

a Predictors: (Constant), Weight (g), Tar (mg), Nicotine content (mg)
b Dependent Variable: Carbon monoxide content (mg)

Table 11.3

The ANOVA F statistic is 78.984, with a p value of .000. From this we can conclude that at least one of the three regression coefficients for our predictors is different from 0 in the population.

Model Summary(b)

Model	R	R Square	Adjusted R Square	Std. Error of the Estimate
1	.958(a)	.919	.907	1.44573

a Predictors: (Constant), Weight (g), Tar (mg), Nicotine content (mg)
b Dependent Variable: Carbon monoxide content (mg)

Table 11.4

When we look at our model summary, shown above in Table 11.4, we find the value of R^2 is .919. This tells us that almost 92% of the variability in carbon monoxide content is explained by linear regression on Weight, Tar and Nicotine.

From the table of coefficients, shown below in Table 11.5, we can derive our fitted regression equation for carbon monoxide content. It is as follows:

Predicted Carbon Monoxide Content = 3.202 + .963(Tar) − 2.632(Nicotine) − .130(Weight)

We can interpret the contribution of each of our explanatory variables in terms of its coefficient. Tar contributes on approximately a 1 for 1 basis. We subtract 2.6 times the Nicotine variable and a small fraction (.13) times the Weight variable. Notice that we interpret the contribution of each of these variables in the context of the other variables. Further discussion is beyond the scope of this text.

Coefficients(a)

Model		Unstandardized Coefficients		Standardized Coefficients	t	Sig.
		B	Std. Error	Beta		
1	(Constant)	3.202	3.462		.925	.365
	Tar (mg)	.963	.242	1.151	3.974	.001
	Nicotine content (mg)	-2.632	3.901	-.197	-.675	.507
	Weight (g)	-.130	3.885	-.002	-.034	.974

a Dependent Variable: Carbon monoxide content (mg)

Table 11.5

Let us do an example. What is the predicted carbon monoxide content for a cigarette with Tar = 1.0, Nicotine = .13, and Weight = .79? To complete the calculations, substitute into the prediction equation above. By substitution we get a predicted carbon monoxide content of $3.202 + .963(1.0) - 2.632(.13) - .130 (.79) = 3.720$.

It is also informative to look at the *t* values, shown in Table 11.5, and their associated *p* values for each of our predictors. Only the *t* value for Tar is significant ($p < .05$). Thus the regression coefficient for this variable is significantly different than 0 while the regression coefficients for our remaining two variables are not, in the context of all three explanatory variables.

Interpretation Of Results

The significance tests noted above seem to contradict the information in Table 11.2 and Figure 11.3. Recall that at that point in the analysis we made a cautionary note about the high intercorrelations among the variables. In other words there is a high degree of overlap in the predictive information that they provide in the multiple regression analysis. Tar adds significantly to our ability to predict carbon monoxide content in cigarettes even after Nicotine and Weight are already in our model.

Residuals

As in simple linear regression, described in Chapter 10, it is important to look at residual plots and look for outliers, influential observations, non-linear relationships, and anything else that stands out. One example is the normal P-P plot shown in Figure 11.8. When the points in this type of plot cluster around a straight line, we have good evidence for a normal distribution. There are many additional plots that can be created and examined for the number of variables in this analysis.

Normal P-P Plot of Regression Standardized Residual

Dependent Variable: Carbon monoxide content (mg)

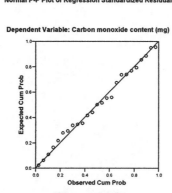

Figure 11.8

Refining The Model

Because Tar has the largest p value of our explanatory variables, and therefore appears to contribute the least to our explanation of carbon monoxide content, we can rerun the regression analysis using only Nicotine and Weight as explanatory variables. The F value from the ANOVA table becomes 66.128 and p value remains at .000. The regression coefficients for this analysis appear in Table 11.6. Our prediction equation for carbon monoxide content now becomes:

Predicted carbon monoxide content = 1.614 + 12.38(Nicotine) + .059(Weight)

Coefficients(a)

Model		Unstandardized Coefficients		Standardized Coefficients	t	Sig.	95% Confidence Interval for B	
		B	Std. Error	Beta			Lower Bound	Upper Bound
1	(Constant)	1.614	4.447		.363	.720	-7.608	10.836
	Nicotine content (mg)	12.388	1.245	.925	9.952	.000	9.807	14.970
	Weight (g)	.059	5.024	.001	.012	.991	-10.360	10.478

a Dependent Variable: Carbon monoxide content (mg)

Table 11.6

As we did earlier, we can interpret the contribution of each of our remaining explanatory variable in terms of its coefficient. Without Tar in the prediction equation, we multiply the Nicotine level by 12.388 and add a small fraction (.059) of the Weight variable. Notice that we interpret the contribution of each of these variables in the context of the other variables. Further, the contributions are dramatically different in this analysis than in our earlier one with all three explanatory variables in the regression analysis. Further discussion is beyond the scope of this text.

Let us return to our sample calculations. What is the predicted carbon monoxide content for a cigarette with Tar = 1.0, Nicotine = .13, and Weight = .79. To complete the calculations, substitute into the prediction equation above. Since Tar is no longer in the equation, we ignore that information and get a predicted carbon monoxide content of 1.614 + 12.38(.13) + .59 (.79) = 3.69. Finally, you might wish to consider rerunning the analysis without Benson and Hedges, which appears to be an outlier in terms of both Tar and Nicotine content.

Exercises For Chapter 11

1. Assessing the worth of a diamond stone is no easy task in view of the four C's, namely carat, color, clarity, and cut. Statistics offers an avenue to infer the pricing of these characteristics. The objective is to infer a sensible pricing model for diamond stones based on data pertaining to their weight (in carats), their color (either D, E, F, G, H, or I), and clarity (either IF, VVS1, VVS2, VS1, or VS2). Of interest is the relative worth of the different grades of color and clarity and whether differences in prices can be attributed to the three different certification bodies (either GIA, IGI, or HRD).

   Additional information about these data can be found in the "Datasets and Stories" article "Pricing the C's of Diamond Stones" in the *Journal of Statistics Education* (Chu, 2001).

   a. Open the data set called Diamond C's using SPSS. It can be accessed through the IPS Web site and is associated with Chapter 11.
   b. Of note in this dataset is the presence of nominal, ordinal, as well as quantitative data. Determine which variable is of which type.
   c. Do a thorough exploratory data analysis including frequencies for each of the four C's (carat, color, clarity, and certification agency) and mean and standard deviation for price.
   d. Explain why it is not meaningful to find means and standard deviations for the four C's. (*Note:* the value labels will give you a hint.)
   e. Following the protocol outlined in the case study at the end of IPS Chapter 11, set up at least one prediction model for Price in Singapore dollars.

2. Baseball provides a rare opportunity to judge the value of an employee — in this case, a player — by standardized measures of performance. The question is, what are those characteristics worth? In addition, the economic principle of freedom of movement for employees can be measured; that is, what financial benefit does a person gain if he is able to change employers? Additionally, baseball fans may use their analysis of this dataset, in combination with other similar datasets, to gain insight into the salary structure in Major League Baseball.

   Consider as our population of interest the set of Major League Baseball players who played at least one game in both the 1991 and 1992 seasons, excluding pitchers. This dataset contains the 1992 salaries for that population, along with performance measures for each player from 1991. Four categorical variables indicate how free each player was to move to other teams.

   Additional information about these data can be found in the "Datasets and Stories" article "Pay for Play: Are Baseball Salaries Based on Performance?" in the *Journal of Statistics Education* (Watnik, 1998).

   a. Open the data set called PayForPlay using SPSS. It can be accessed through the IPS Web site and is associated with Chapter 11.
   b. Of note in this dataset is the presence of various data types. Determine the type of variable (nominal, ordinal, quantitative, etc.) for each variable in the data set.
   c. Do a thorough exploratory data analysis including visual and numerical summaries for each variable and for the relationships between the variables.
   d. Where the data indicate it is required, do a transformation on the response variable.
   e. Detect and eliminate outliers as appropriate.
   f. Following the protocol outlined in the case study at the end of IPS Chapter 11, set up at least one prediction model for Salary and interpret the coefficients.

Chapter 12. One-Way Analysis Of Variance

Topics Covered In This Chapter:

12.1 Inference For One-Way Analysis of Variance
12.2 Comparing The Means

This chapter describes how to perform **one-way ANOVA** using SPSS for Windows for determining whether the means from several populations differ. Specifically, the null and alternative hypotheses for one-way ANOVA are H_0: $\mu_1 = \mu_2 = \ldots = \mu_I$ and H_a: not all of the μ_i are equal.
If the overall F value is significant, then follow-up (multiple) comparisons are conducted.

12.1 Inference For One-Way Analysis of Variance

The example used in IPS for illustrating ANOVA calculations is described in Example 12.3. The example is about the Safety Climate Index (SCI) that was obtained from Unskilled Workers, Skilled Workers, and Supervisors. Figure 12.3 in IPS gives us the means, standard deviations, and sample sizes for the data. We are not given the full dataset. Using the information contained in Figure 12.3 in IPS and the protocol developed in Chapter 3 of this manual, I have generated random samples taken from each of the three populations described. In what follows, the same procedures are followed as in IPS, however, because we are working with different samples from the same population, the results shown here will not match exactly the IPS outcomes. However, given the basic logic of statistics, 95% of the time we should arrive at the same decision based on our F values at the end of the process.

Before proceeding with ANOVA, we must verify that the assumptions of normally distributed data and equality of standard deviations in the various groups are reasonably satisfied. Histograms with a normal curve over them are shown in Figure 12.1. The assumption of normality appears to be met.

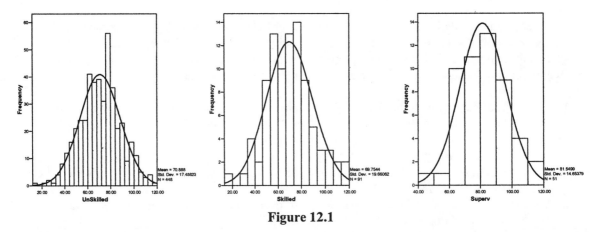

Figure 12.1

The descriptive statistics are shown in Table 12.1 on the following page. A scan of the standard deviations shows that our largest standard deviation (19.66) is not more than twice as large as the smallest (14.65). Thus by our "rule of thumb" the assumption of equality of variances has been satisfactorily met and we can proceed with our ANOVA.

Descriptive Statistics

	N	Minimum	Maximum	Mean	Std. Deviation
UnSkilled	448	14.46	117.64	70.8880	17.48823
Skilled	91	14.44	119.49	69.7544	19.66062
Superv	51	45.57	113.78	81.5499	14.65379
Valid N (listwise)	51				

Table 12.1

Now we must set up our data in a format that will work for SPSS. This requires that the data for all three groups be stacked on one another into a single column. Beside that column, place the number 1 beside the first 448 entries and label that as the Unskilled Group in the "Values" column on the "Variable View" page. Continue to stack the SCI scores for each of the other groups until you have only two columns. A portion of the dataset is shown below in Figure 12.2.

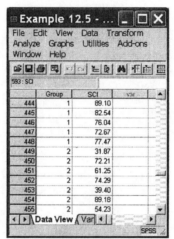

Figure 12.2

To perform a one-way ANOVA with post hoc multiple comparisons, follow these steps:

1. Click **Analyze,** click **Compare Means,** and then click **One-Way ANOVA.** The "One-Way ANOVA" window in Figure 12.3 appears (see following page).
2. Click *SCI,* then click ▸ to move *SCI* into the "Dependent List" box.
3. Click *Group,* then click ▸ to move *Group* into the "Factor" box.
4. If you are interested in multiple comparisons, you can click on the **Post Hoc** box. The SPSS window in Figure 12.4 appears (see following page). Click on the desired multiple comparison test procedure (e.g., LSD).
5. Click **Continue.**
6. Click **Options** and you will get a screen like that shown in Figure 12.5. Click on "Descriptives" and "Means Plot. Click **Continue.**
7. Click **OK.**

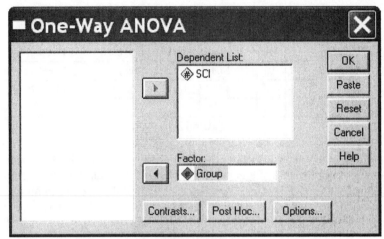

Figure 12.3

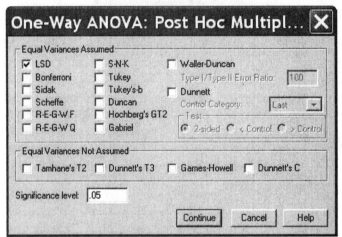

Figure 12.4

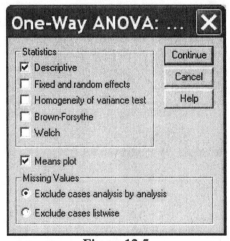

Figure 12.5

The ANOVA summary table (Table 12.2), the LSD method for multiple comparison (Table 12.3), and a means plot (Figure 12.5) from the SPSS output are included on the next page.

ANOVA

SCI

	Sum of Squares	df	Mean Square	F	Sig.
Between Groups	5585.404	2	2792.702	8.996	.000
Within Groups	182234.936	587	310.451		
Total	187820.340	589			

Table 12.2

Because the *p*-value ("Sig." in table 12.2) is less than 0.001 there is sufficient evidence to reject H_0 in favor of H_a. This indicates that some μ_i differ. An initial comparison for the SCI ratings for the three types of employees can be made from the descriptive statistics in Table 12.1 earlier. Recall that our samples differ from those used in IPS, however, the population from which they were drawn is the same (we used the same means and standard deviations as the original). Note also that we arrived at the same conclusion in terms of rejecting the null hypothesis.

12.2 Comparing The Means

In many studies specific questions cannot be formulated in advance. If H_0 is rejected, we would like to know which means differ from one another. One procedure that is commonly used for multiple comparisons is the least-significant differences (LSD) method. The output of the LSD method, found in Table 12.3, indicates significantly different means with an asterisk (*). The LSD method indicates that the means for *UnSkilled* and *Skilled* do not differ significantly. The differences between the SCI scores for *UnSkilled* and *Superv* and for *Skilled* and *Superv* do differ significantly. Here again, we have not used exactly the same procedure as illustrated in your text, however, we have reached the same conclusions about the specific differences between the SCI scores for each of the employee groups.

Multiple Comparisons

Dependent Variable: SCI
LSD

(I) Group	(J) Group	Mean Difference (I-J)	Std. Error	Sig.	95% Confidence Interval	
					Lower Bound	Upper Bound
UnSkilled	Skilled	1.13358	2.02596	.576	-2.8454	5.1126
	Superv	-10.66192(*)	2.60389	.000	-15.7760	-5.5478
Skilled	UnSkilled	-1.13358	2.02596	.576	-5.1126	2.8454
	Superv	-11.79550(*)	3.08202	.000	-17.8486	-5.7424
Superv	UnSkilled	10.66192(*)	2.60389	.000	5.5478	15.7760
	Skilled	11.79550(*)	3.08202	.000	5.7424	17.8486

* The mean difference is significant at the .05 level.

Table 12.3

The means plot, in Figure 12.6 on the following page, is useful in demonstrating the relationships between the various means. A visual inspection of the plot confirms our analysis that the SCI scores for Skilled and Unskilled Workers do not differ from each other but they both differ from the SCI scores of the Supervisors.

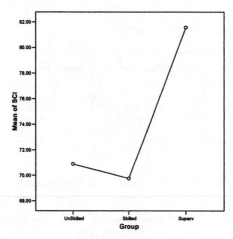

Figure 12.6

Exercises For Chapter 12

1. Explain what is meant by this statement that I made earlier in this chapter:
    [T]he same procedures are followed as in IPS, however, because we are working with different
        samples from the same population, the results shown here will not match exactly the IPS
        outcomes. However, given the basic logic of statistics, 95% of the time we should arrive at the
        same decision based on our F values at the end of the process.

2. The following data are for four independent groups:
    a. Do a complete exploratory data analysis.
    b. Is there a significant difference between the four groups?
    c. If there is a significant difference, complete a multiple comparisons analysis to determine which
        group(s) are different from other group(s).

Group 1	Group 2	Group 3	Group 4
11.61	9.78	14.40	14.65
14.37	6.59	14.27	12.28
13.61	15.14	11.13	12.08
13.35	8.78	15.46	5.37
8.62	9.68	13.53	14.30
8.90	14.59	11.81	5.84
13.48	8.97	12.64	7.21
19.64	8.59	14.52	19.13
13.89	12.21	14.37	21.04
13.48	6.23	9.61	13.80
13.41	9.72	15.38	19.18
10.82	13.50	12.14	26.59
18.99	4.79	13.50	17.44
11.73	8.73	12.68	21.21
13.07	6.95	10.69	9.42
8.85	8.61	14.12	11.11
9.67	5.67	12.09	11.13
17.68	13.82	11.45	15.72
16.80	14.56	9.81	18.49
16.33	5.37	13.96	9.81

3. Using the means and standard deviations for the dataset given in Question 2 above, generate another
    simple random sample from the same population and repeat the steps for Question 2. In addition,
    explain why the outcome is the same or different from your answer to Question 2. Your outcome for
    Question 2 should match exactly with that of your classmates. Will your outcome for Question 3
    match exactly? Why or why not?

4. A scientist investigates the neural pathways involved in memory. To complete the investigation, 32
    rats are trained to run a maze until they complete five errorless runs in a row. The animals are then
    randomly assigned to one of four groups. Subjects are anesthetized and then receive an electrically
    induced brain lesion in one site, a second site, in both sites, or receive sham surgery. In the sham
    surgery, all preparations for lesioning are made, including placing an electrode in the brain, but no
    electrical current is used. After a recovery period, the rats are retested in the maze. The data on the
    next page show the number of trials to relearn the maze to the original criterion of five errorless runs
    in a row.
    a. Do a complete exploratory data analysis.
    b. Is there a significant difference between the four groups?
    c. If there is a significant difference, complete a multiple comparisons analysis to determine which
        group(s) are different from other group(s).

Site 1	Site 2	Site 1 + 2	Sham Surgery
2	4	5	3
3	3	11	6
3	3	10	3
3	2	10	2
1	7	9	4
1	2	7	1
3	3	12	3
6	1	8	5

5. From the data below, complete an analysis of variance where $\alpha = .01$.
    a. Do a complete exploratory data analysis.
    b. Is there a significant difference between the four groups?
    c. If there is a significant difference, complete a multiple comparisons analysis to determine which group(s) are different from other group(s).

Group 1	Group 2	Group 3	Group 4
2	6	2	7
3	4	7	8
3	5	2	8
3	7	2	4
2	3	5	9
2	5	5	6
3	4	1	8
5	5	6	7
3	3	4	5
3	7	2	8

Chapter 13. Two-Way Analysis of Variance

Topics covered in this chapter:

13.1 The Two-Way ANOVA Model

In a one-way ANOVA, we deal with one independent variable, or factor. In a two-way ANOVA, we deal with two independent variables, or factors. The considerations in this chapter can be applied to other study designs involving more than two independent variables, however, the focus will be on the two-way ANOVA model.

ANOVA designs are generally referred to in terms of the number of factors (two in this chapter) and the number of levels in each factor. Example 13.1 is a 3 * 3 design. There are two factors (therefore two numbers) and each factor has three levels, therefore it is a 3 * 3 design. Example 13.4 is a 3 * 2 design, and Example 13.6 is a 2 * 2 design. You can confirm this by your own reading of these examples.

Advantages Of Two-Way ANOVA

The main difference between one-way and two-way ANOVA is in the FIT part of the model. By combining two experiments (for example, the one discussed in Examples 13.1 and 13.2 of IPS), we can increase our efficiency by examining two factors at once; investigate possible interactions between the factors in our study; and reduce the residual variation by including a second factor we think might influence the outcome or response of another factor. Further illustrations are included in Examples 13.3 to 13.6.

The Two-Way ANOVA Model

As stated earlier, there are two factors in a two-way ANOVA. Each factor has its own number of levels. Example 13.2 in IPS shows a table that displays a two-way design for the information discussed in Example 13.1 of IPS. The two factors are logo and color and each factor has three levels, which makes this a 3 * 3 design (three types of logos * three choices of color). In a two-way design, every level of one factor is paired with every level of the other factor so that all combinations are compared.

The underlying assumptions of the two-way ANOVA model are that we have independent SRS's from each population.

Main Effects And Interactions

To illustrate the application of SPSS to calculate main effects and interactions, we will use Example 13.7 from IPS.

To begin, enter the data into an SPSS spreadsheet using three columns. In the first two columns we show the levels of the two factors, *Year* and *Location*. Column 3 shows the number of *calories* per portion in soft drinks consumed. See Figure 13.1.

	Year	Location	Calories	var	var	var	var	v
1	1978	1	130					
2	1978	2	125					
3	1991	1	133					
4	1991	2	126					
5	1996	1	158					
6	1996	2	155					
7								
8								

Figure 13.1

Begin your analysis by completing an exploratory data analysis (EDA). Use the following steps as a guide:

1. Split the file by Year and complete an EDA. See Figures 13.2, 13.3 and 13.4.
2. Split the file by Location and complete an EDA. See Figure 13.4.
3. Ungroup the variables (reverse the last split file command) and create a clustered bar graph. See Figure 13.5. This last figure gives the same information as Figure 13.1 in IPS. The production of the equivalent line graphs will be explored in section 13.2 of this chapter.
4. Write a brief summary about the means and interaction effects that you see in these figures.

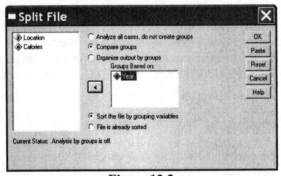

Figure 13.2

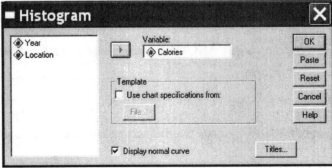

Figure 13.3

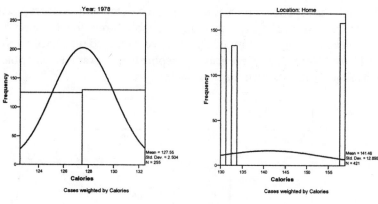

Figure 13.4

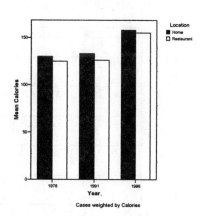

Figure 13.5

We can repeat this process for Examples 13.8 to 13.10 as well. The clustered bar graphs are shown in Figure 13.6 below. Write the summaries for these examples as well. See the text summary and Figures 13.2 to 13.4 in IPS for comparison.

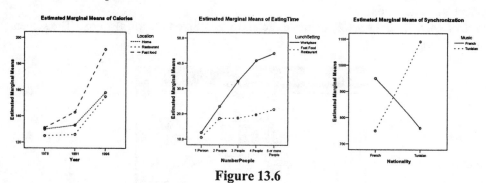

Figure 13.6

13.2 Inference For Two-Way ANOVA

To illustrate the ANOVA calculations using SPSS, refer to the following example. In order to test the separate and mutual effects of two drugs, A and B, on physiological arousal, researchers randomly and independently assigned 40 laboratory rats to 4 groups of 10 subjects each. Each group received zero units or 1 unit of Drug A and zero units or 1 unit of Drug B. The dependent variable was a standard measure of physiological arousal. One of the groups served as a control, receiving only an inert placebo containing zero units of Drug A and zero units of Drug B. The following table shows the measures of physiological arousal for each subject in each of the four groups. The data were entered into an SPSS spreadsheet with three variables labeled *UnitsA, UnitsB,* and *Arousal.* To replicate this example, enter these data into your own spreadsheet and then follow the instructions given below.

Drug A		Drug B	
		0 units	1 unit
	0 units	20.4 17.4	20.5 26.3
		20.0 18.4	26.6 19.8
		24.5 21.0	25.4 28.2
		19.7 22.3	22.6 23.7
		17.3 23.3	22.5 22.6
	1 unit	22.4 19.1	34.1 21.9
		22.4 25.4	32.6 28.5
		26.2 25.1	29.0 25.8
		28.8 21.8	29.0 27.1
		26.3 25.2	25.7 24.4

1.   Complete an EDA as shown above, then reverse the **Split File** action because we want all of the data included in the ANOVA calculations.
2.   Click on **Analyze, General Linear Model, Univariate.** Click on *Arousal* and move it into the "Dependent Variable:" box.
3.   Click on *UnitsA,* and then *UnitsB,* to move each into the "Fixed Factor(s):" box. The screen shown in Figure 13.7 appears.

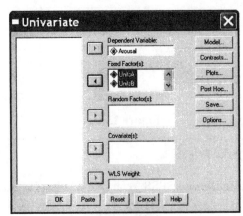

Figure 13.7

4. Click on Plots.
5. Click on *UnitsA* to move it into the Horizontal Axis box.
6. Click on *UnitsB* to move it into the "Separate Lines" box. See Figure 13.8 on the following page.

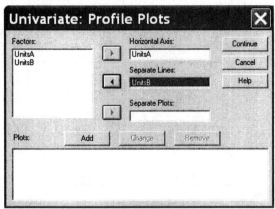

Figure 13.8

7. Click **Add.** Click **Continue.**
8. Click **Model.** Click the ✓ beside "Include intercept in model" to turn this feature off. See Figure 13.9.
9. Click **Continue.**
10. Click **OK.**

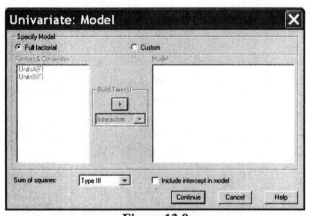

Figure 13.9

The ANOVA Table For Two-Way ANOVA

The ANOVA summary table is shown on the following page in Table 13.1. Note that there is a significant effect of the number of Units of Drug A ($F = 17.7$, $p = .000$), a significant effect of the number of Units of Drug B ($F = 13.8$, $p = .001$), but no interaction effect ($F = .01$, $p = .936$). These effects are illustrated in Figure 13.10 on the next page. Review the plots and your text to ensure that you understand how to read the summary table and the plot of the means.

Tests of Between-Subjects Effects

Dependent Variable: Arousal

Source	Type III Sum of Squares	df	Mean Square	F	Sig.
Model	23472.063(a)	4	5868.016	676.413	.000
UnitsA	153.272	1	153.272	17.668	.000
UnitsB	120.062	1	120.062	13.840	.001
UnitsA * UnitsB	.056	1	.056	.006	.936
Error	312.307	36	8.675		
Total	23784.370	40			

a R Squared = .987 (Adjusted R Squared = .985)

Table 13.1

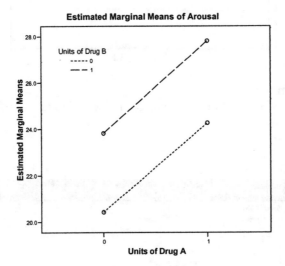

Figure 13.10

Exercises For Chapter 13

1. Consider a hypothetical experiment on the effects of a stimulant drug on the ability to solve problems. There were three levels of drug dosage: 0 mg, 100 mg, and 200 mg. A second variable, type of task, was also manipulated. There were two types of tasks: a simple well-learned task (naming colors) and a more complex task (finding hidden figures in a complex display). The mean time to complete the task for each condition in the experiment is shown below.
    a. Use the techniques of EDA to generate means, standard deviations, and to check that the standard deviations are approximately the same and that the data are normally distributed.
    b. Create the ANOVA summary table and plots of the means.
    c. Summarize the findings in a short paragraph.

0 mg		100 mg		200 mg	
Simple	Complex	Simple	Complex	Simple	Complex
18.3	79.9	19.9	90.5	21.2	77.1
31.4	95.9	7.6	97.0	7.8	59.1
27.9	59.2	46.4	91.4	43.8	106.9
26.6	75.8	11.9	97.1	17.0	111.1
4.2	68.3	18.3	92.9	20.8	101.1
5.5	75.3	11.8	86.2	41.5	80.4
27.2	62.9	2.1	103.4	17.8	88.2
56.5	97.2	40.2	92.0	16.2	130.2
29.2	100.3	36.0	91.9	31.4	84.9
27.2	61.7	33.8	95.6	6.3	87.7

2. The consumption of alcohol has been shown to interfere with learning and may impair the developing brain permanently. In an experimental study, groups of rats were fed either normal water in their water bottles or a 35% solution of alcohol either at adolescence or adulthood. These rats were then trained to run in a maze for food. The number of trials it took to run the maze perfectly three times in a row was recorded. The higher the number of trials, the poorer the performance. The data are shown below.
    a. Use the techniques of EDA to generate means, standard deviations, and to check that the standard deviations are approximately the same and that the data are normally distributed.
    b. Create the ANOVA summary table and plots of the means.
    c. Summarize the findings in a short paragraph.

Adolescent		Adult	
0% Alcohol	35% Alcohol	0% Alcohol	35% Alcohol
5	18	6	6
4	19	7	9
3	14	5	5
4	12	8	9
2	15	4	3

3. We are examining the results of a study of eyewitness testimony. The investigators wish to know whether recall of important information is influenced by the severity of the crime and the amount of time that passes before the witness is asked to recall important events. A videotape is made of a man stealing a woman's purse in an uncrowded clothing store. A second videotape is made, using the same actors, of a man using a gun to rob the cash register of the same clothing store. In this scenario, the woman is a sales clerk behind the cash register and there are two people standing at the counter. Subjects are asked to view the videotape and to imagine that they are at the crime scene. In the "immediate" condition, subjects are asked a series of questions about the crime within 10 minutes of

viewing the videotape. In the "delay" condition, they are asked the same series of questions about the crime 20–30 minutes after viewing the tape. This is designed to be similar to an actual crime scene where there is some delay when taking witness statements. The dependent variable is the total number of questions answered correctly. Subjects were college students who were randomly assigned to experimental groups; 10 subjects were assigned to each group. The data are shown in the table below.

a. Use the techniques of EDA to generate means, standard deviations, and to check that the standard deviations are approximately the same and that the data are normally distributed.
b. Create the ANOVA summary table and plots of the means.
c. Summarize the findings in a short paragraph.

	Immediate Recall	Delayed Recall
Armed Robbery	7	3
	5	4
	7	3
	7	4
	6	5
	6	4
	7	3
	5	3
	6	5
	7	4
Purse Snatching	7	5
	7	5
	8	5
	8	7
	8	7
	8	6
	7	6
	8	5
	9	5
	7	4

4. Let's assume we're planting corn. The type of seed and type of fertilizer are the two factors we're considering in this example. The data that actually appear in the table are samples. In this case, two samples from each treatment group were taken.
a. Use the techniques of EDA to generate means, standard deviations, and to check that the standard deviations are approximately the same and that the data are normally distributed.
b. Create the ANOVA summary table and plots of the means.
c. Summarize the findings in a short paragraph.

	Fert 1	Fert 2	Fert 3	Fert 4	Fert 5
Seed A	106, 110	95, 100	94, 107	103, 104	100, 102
Seed B	110, 112	98, 99	100, 101	108, 112	105, 107
Seed C	94, 97	86, 87	98, 99	99, 101	94, 98

5. An evaluation of a new coating applied to three different materials was conducted at two different laboratories. Each laboratory tested three samples from each of the treated materials. The results are given in the table below.
a. Use the techniques of EDA to generate means, standard deviations, and to check that the standard deviations are approximately the same and that the data are normally distributed.
b. Create the ANOVA summary table and plots of the means.
c. Summarize the findings in a short paragraph.

	Materials		
	1	2	3
Lab 1	4.1	3.1	3.5
	3.9	2.8	3.2
	4.3	3.3	3.6
Lab 2	2.7	1.9	2.7
	3.1	2.2	2.3
	2.6	2.3	2.5

6. Suppose a statistics teacher gave an essay final to her class. She randomly divides the classes in half such that half the class writes the final with a standard examination book and half with notebook computers. In addition the students are partitioned into three groups, low Anxiety, medium Anxiety, and high Anxiety. Answers written in exambooks will be transcribed to word processors and scoring will be done blindly. That is, the instructor will not know the method or Anxiety level of the student when scoring the final. The dependent measure will be the score on the essay part of the final exam. The data are shown below.
   a. Use the techniques of EDA to generate means, standard deviations, and to check that the standard deviations are approximately the same and that the data are normally distributed.
   b. Create the ANOVA summary table and plots of the means.
   c. Summarize the findings in a short paragraph.

Method	Anxiety	Score
Standard Exam Book	High	23
	High	32
	High	25
	Medium	29
	Medium	30
	Medium	34
	Low	31
	Low	36
	Low	33
Notebook Computer	High	32
	High	26
	High	26
	Medium	34
	Medium	41
	Medium	35
	Low	23
	Low	26
	Low	32

7. A computer magazine recently conducted a comparison of four inkjet printers. A page containing a picture and some text was printed using two types of paper (rag and glossy) with each printer. Two copies were made for each printer and paper combination. These were then shown in random order to a panel of judges who rated the overall quality of each print on a 1–10 scale, with 10 being perfect. The average rating for each print was used as the response variable. The results are shown on the next page.
   a. Use the techniques of EDA to generate means, standard deviations, and to check that the standard deviations are approximately the same and that the data are normally distributed.
   b. Create the ANOVA summary table and plots of the means.
   c. Summarize the findings in a short paragraph.

	Printer			
	HP	Epson	Lexmark	NEC
Rag	8, 9	7, 6	7, 7	4, 4
Glossy	6, 5	8, 9	7, 9	9, 7

8. A clinical psychologist is investigating the effect of a new drug combined with therapy on schizophrenic patients' behavior. The drug has three dosages given (absent, low, high) and the therapy has four types (behavior modification, B; psychodynamic PD; group counseling, Grp; nondirective counseling, ND). Higher scores indicate greater positive effects.
   a. Use the techniques of EDA to generate means, standard deviations, and to check that the standard deviations are approximately the same and that the data are normally distributed.
   b. Create the ANOVA summary table and plots of the means.
   c. Summarize the findings in a short paragraph.

Absent				Low				High			
BM	PD	Grp	ND	BM	PD	Grp	ND	BM	PD	Grp	ND
25	27	20	19	25	25	22	23	22	28	27	20
22	25	25	22	21	23	21	27	30	27	25	24
23	26	19	20	20	19	25	23	26	24	28	21
25	24	21	17	22	24	26	26	28	21	29	24
24	25	18	22	21	20	22	28	20	25	24	26

9. Two dose levels of a neurotoxic drug were given to rats at three age levels. In adulthood, the subjects were tested for balance and coordination. Higher scores indicate poor balance and coordination.
   a. Use the techniques of EDA to generate means, standard deviations, and to check that the standard deviations are approximately the same and that the data are normally distributed.
   b. Create the ANOVA summary table and plots of the means.
   c. Summarize the findings in a short paragraph.

High Dose			Low Dose		
Age in Days			Age in Days		
10	20	40	10	20	40
12	15	27	12	32	8
12	23	22	11	24	11
8	15	22	5	18	9
14	16	22	15	22	11
11	27	24	9	28	12
9	29	26	7	23	1
10	22	22	17	18	9
12	27	30	13	23	11
12	33	23	10	26	10

10. In thinking about the Obedience to Authority study, many people have thought that women would react differently than men. Some have said that because women were more willing to obey others, they would shock at higher levels. Others have stated that because women as a group are so opposed to violence they would quit the experiment earlier than men. Still others have contended that there would be a major interaction between the situation in which the people were placed, and their gender. They believed that women in the Remote situation would break off the experiment early; they thought that they would shock at higher levels in the Voice and In-Room situations, but would again stop at lower shock levels when they had to touch the learner. A different pattern was predicted for men. Data like that collected in the Obedience study is reproduced below.
   a. Use the techniques of EDA to generate means, standard deviations, and to check that the standard deviations are approximately the same and that the data are normally distributed.

b. Create the ANOVA summary table and plots of the means.
c. Summarize the findings in a short paragraph.

Remote		Voice		In-Room		Must-Touch	
Gender	Volts	Gender	Volts	Gender	Volts	Gender	Volts
M	300	M	135	M	105	M	135
F	300	F	150	F	150	F	150
M	315	M	150	M	150	M	150
F	315	F	165	F	150	F	150
M	330	M	285	M	150	M	150
F	345	F	315	F	150	F	150
M	375	M	315	M	180	M	150
F	450	F	360	F	270	F	150
M	450	M	450	M	300	M	150
F	450	F	150	F	300	F	180
M	450	M	450	M	315	M	210
F	450	F	450	F	315	F	255
M	450	M	450	M	450	M	300
F	450	F	450	F	450	F	315
M	450	M	450	M	450	M	450
F	450	F	450	F	450	F	450
M	450	M	450	M	450	M	450
F	450	F	450	F	450	F	450
M	450	M	450	M	450	M	450
F	450	F	450	F	450	F	450

Chapter 14. Bootstrap Methods and Permutation Tests[1]

Topics covered in this chapter:

This chapter introduces the concepts of bootstrapping and permutation tests — new methods that use large numbers of relatively simple computations to produce confidence intervals and tests of significance in situations in which the data do not meet the assumptions for traditional statistical tests. For example, when the data are strongly skewed, we cannot use *t* tests and must look for an alternative. Such alternatives are introduced here.

14.1 The Bootstrap Idea

When we complete tests of statistical inference, we are basing our decisions on sampling distributions of sample statistics. The bootstrap, on the other hand, provides a way of finding the sampling distribution from a single sample. Bootstrapping has two overlapping purposes. First, it permits us to deal with distributions that are not normal. Second, and more important, it allows us to estimate parameters that we do not know how to estimate by the use of a formula.

Think of examples we have used thus far in the text. When we have normal distributions, we have formulas for calculating the mean, standard deviation, standard error, and confidence intervals. But, what do we do if we wish to calculate confidence limits for our estimates with non-linear regression, for example? Bootstrapping is a relatively new means of overcoming the need for normally distributed data and it is used widely in today's world of data mining.

Throughout this chapter we will use SPSS to illustrate the methods of bootstrapping. Alternatively, follow the directions in the text for downloading S-Plus. Sample SPSS macros are posted at www.whfreeman.com/ips5e.

---

[1] I acknowledge previous SPSS programmers including David Marso, Rolf Kjoeller, Raynald Levesque and others who have generously posted their bootstrapping macros on the web. I also acknowledge David Howell and our text authors for their dedication to statistical solutions and excellence in teaching.

In bootstrapping, the calculations are relatively simple and must be repeated many times over. This is the perfect situation for developing a macro, a set of commands that can be run repeatedly in a semi-automatic way. I have begun the process of creating macros to do bootstrapping. While fully functional, they lack a certain elegance. Writing macros is not difficult but the debugging process can be challenging for a new programmer like me.

To begin, download the data for Example 14.1 from the IPS Web site and save it in an SPSS file called Example 14.1.sav. Label the variable as *Time*. Replicate Figure 14.1 in IPS using the SPSS techniques that you already know. The results are shown below in Figure 14.1. Check these against Figure 14.1 in IPS.

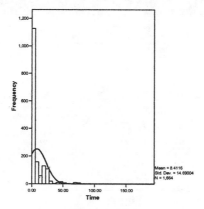

Figure 14.1

The first step in bootstrapping is to sample with replacement until you have a sample of the same size as your original sample. You can create your own macro for doing this using the following steps.

1. Open a blank SPSS spreadsheet.
2. Click **Edit** and then **Options.**
3. Click beside "Record syntax in journal." See Figure 14.2.
4. Click **File, Open, and Data** and open "Example 14.1.sav."

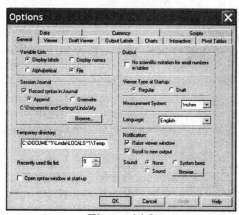

Figure 14.2

5. With the file open, click **Data, Select Cases,** "Random Sample of cases."
6. Click "Sample," "Exactly" 1 "of the first" 1664 "cases." Click **Continue.** Click **OK.**
7. Click **Data, Sort Cases** by "filter," "descending." Click **OK.**

8. Click **Transform, Compute.** Enter a "Target variable" name such as 'Sample.'
9. Double-click on "Time" so that it moves into the "Numeric expression" box.
10. Now click "If…" below the number pad in the "Compute Variable" window. The window shown in Figure 14.3 appears.
11. Click "Include if case satisfies condition."
12. Double-click "filter_$" to move it into the blank window. Now add "= 1" so that the expression reads "filter_$=1".
13. Click **Continue.** Click **OK.**
14. Click **File, Save.**

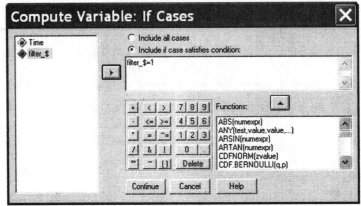

Figure 14.3

Now click on the "Output-1 SPSS Viewer Window." The window will contain the code required to re-run the process you have just completed. See Figure 14.4.

1. Select all the entries in the output window and copy the contents.
2. Click **File, New, Syntax.** See Figure 14.5.
3. Click in the blank window and then right-click "Paste" to enter the syntax as copied from the output window.
4. Click **File, Save As** and give the macro a name. Mine is called "Sampling With Replacement." Be sure that the file type is shown as "SPSS Syntax files." See Figure 14.6.
5. Click **Run, All** as shown in Figure 14.7 on the following page. Notice that you now have a second entry in the variable named *Sample.* Repeat this step as often as you have patience to do it, preferably 1664 times! Notice that each run adds another selection to the variable called *Sample.*

Figure 14.4

Figure 14.5

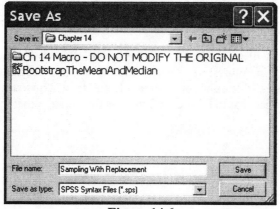

Figure 14.6

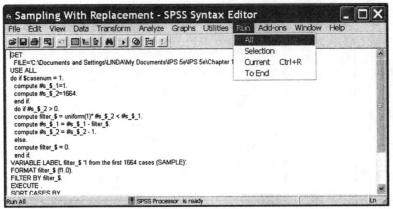

Figure 14.7

Our objective in bootstrapping is to first get a new sample of 1664 entries, sampled with replacement, from our original data set. To do so means repeating Step 5 above 1664 times! This completes Step 1 of the process. Then, at Step 2, we want to repeat that entire process at least 1000 times so that we have 1000 samples like the first one. For each of our resamples we calculate a statistic of interest and save it (the mean of each sample, for example). This is an ideal task for a computer since it involves such a large number of replications.

Go to www.whfreeman.com/ips5e and you will find a macro called "BootstrapTheMeanAndMedian". Download it and then run it using the instructions given below. This macro will open the data file called "Ex 14.1.sav", run Steps 1 and 2 of the bootstrapping process, and save the output to a file called "EX14.1_OUTPUT1." The macro can be adapted for any data set by inserting the appropriate sample size, number of replications, the appropriate file name to access the data, and by specifying output file names that are meaningful to you.

To open and run the macro, follow these steps:
1. Double-click on the macro "BootstrapTheMeanAndMedian" to open it.
2. Click **Run, All.**
3. It is necessary to run the syntax twice – once to initialize the variables and then once to run the program.
4. As the program runs, you will see information in the lower left "information area" of the SPSS Viewer page as illustrated in Figure 14.8. Do not interrupt the processing until you see the message "SPSS processor is ready" in this information area.   Better to wait an extra few seconds than to interrupt the process! Notice that in Figure 14.8 there is the notation that there are "Transformations pending."

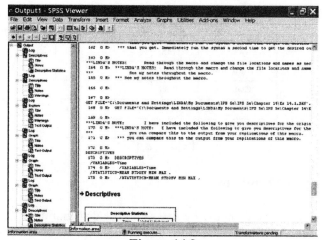

Figure 14.8

When you run the macro, you will get an SPSS spreadsheet called "EX14.1_OUTPUT1.sav." The macro computes the mean and median as well as a bias measure that is calculated as the difference between the mean of the original data set and the mean of the sample for each of the 1000 samples. The macro then computes descriptive statistics on all three measures. These descriptive statistics and graphs give us the bootstrap distribution. The bootstrap distribution represents the sampling distribution of the statistic based on many resamples.

Figure 14.9 shows a distribution for one pass through the macro described above. Compare this to Figure 14.3 in IPS. The distribution of sample means is nearly normal. The mean of the means for these 1000 resamples from the original data set is 8.4171 which is very close to the mean of the original data set (8.41). The shape of the distribution is nearly normal. The bootstrap standard error is the standard deviation of the bootstrap distribution and for this particular resample it is .36117 which is also very close to the theory-based estimate of the standard error based on the original sample. Notice that SPSS prints these values beside the histogram and you can read them from Figure 14.9 below and Table 14.1 on the following page.

The bootstrap distribution of means, as shown in the figure below, is used to estimate how the sample mean of one actual sample of size 1664 would vary because of random sampling. By adding the parameter estimates as we did in the paragraph above, we are actually applying the "plug-in" principle. That is, we substitute the data for the population.

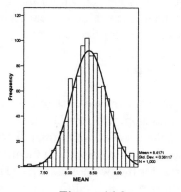

Figure 14.9

Descriptive Statistics

	MEAN	Valid N (listwise)
N	1000	1000
Minimum	7.13	
Maximum	9.41	
Mean	8.4171	
Std. Deviation	.36117	

Table 14.1

14.2 First Steps In Using The Bootstrap

The bootstrap is very useful when we do not know the sampling distribution of the statistic of interest. Because the shape of the bootstrap distribution closely approximates the shape of the sampling distribution, we can use it to check the normality of the sampling distribution. Further, we can check to see whether the bootstrap distribution is centered at the value of the statistic for the original sample. That is, we can compute a measure of bias that reflects the difference between the mean of the bootstrap distribution and the value of the same statistic in the original sample.

Bootstrap T Confidence Intervals

Example 14.4 and Table 14.1 in IPS show the selling prices of a random sample of 50 properties sold in Seattle during 2002. The sample includes both residential and commercial properties. For this example in IPS, we are asked to use the 25% trimmed mean; that is, the mean of the middle 50% of the sample. By doing so, we eliminate both the least and the most expensive properties. This statistic may be more representative of property prices than the median value.

We can replicate our *t* test for a single sample using a measure of the center of the data set and the bootstrapping protocol developed for bootstrapping the median. Since the bootstrap syntax included here does not include a measure for the 25% trimmed mean of the sample we will use the median (see "BootstrapTheMeanAndMedian") for the examples in this section. Check the Web site at www.whfreeman.com/ips5e for other macros.

First, open the macro and modify the file names (input and output) as appropriate. Run the syntax and view the output provided (descriptive statistics and histograms). Inspect the histogram for the bootstrapped median to determine if it is approximately normal in shape. Locate the bootstrap standard error of the median. Calculate the 95% confidence interval for the median. From our bootstrap distribution, we are 95% confident that the median value for real estate sales in Seattle in 2002 is between the values for the lower and upper limits of the confidence interval.

Bootstrapping To Compare Two Groups

Recall from Chapter 7 that in two-sample problems we wish to compare the mean of one sample to the mean of another based on separate samples from two populations. The bootstrap can also be used to compare two populations without facing the necessity of normal distributions in each population. Further, using the bootstrap approach, we are not limited to comparing the means of the two populations. Rather, we can compare any parameter of interest for the two populations. The permutation test macros referenced in the last section of this chapter may be modified for sampling with replacement and applied here.

First, we must generate independent SRS's from each population. Next, we calculate the statistic of interest for each resample, in this case the difference between the means of the two groups. Finally, we construct the bootstrap distribution for this statistic calculated repeatedly. The principle is the same as it was for bootstrapping the mean. The specific difference is that we are now interested in the bootstrap distribution (and its related bias measure) for the difference between the two means at each stage of the resampling. Check the IPS Web site at www.whfreeman.com/ips5e for available SPSS macros to download and run and/or modify appropriately.

Beyond The Basics: The Bootstrap For A Scatterplot Smoother

We are not limited to bootstrapping traditional parameters such as the mean and standard deviation. We can also bootstrap, for example, information about regression lines. The basic approach is the same as used in all previous examples in this chapter. First, generate a sample with replacement from the original data. Then calculate and plot the regression line. Now repeat this process hundreds of times to get a "smooth" regression line. Any pattern that occurs consistently is considered a real pattern, not one that occurred just by chance.

14.3 How Accurate Is A Bootstrap Distribution?

We can rely on the bootstrap distribution of a relatively large sample from a population to inform us about the shape, bias, and spread of the sampling distribution. How can we be so certain given that there are two sources of variation in any bootstrap distribution? The first source of variation stems from the fact that the sample was first taken randomly from a population and such sampling produces variation from sample to sample. In addition, when we resample, we again add variability due to random sampling from the sample. To illustrate the truth of our opening statement, follow the process shown in Figure 14.12 in IPS.

First, we need a population of values from which to draw our sample. Take a hypothetical case of the length of time required to retrieve parts from a warehouse. The data were generated using the rv.normal function by creating two variables, Time 1 and Time 2 with n = 8000 for each. For Time 1 the mean and standard deviation were 2.8 and 1.5, respectively. For Time 2 these same values were 8 and 1.5. To create the bimodal distribution, these two variables were combined into one called Time with the label Time(min). You can create a similar data set to follow this example if you choose. A histogram of time taken to retrieve these parts is shown on the following page in Figure 14.10. Compare this distribution to the one shown in the upper left of Figure 14.12 in IPS. What follows is a replication of the steps required to obtain the output shown in the remainder of Figure 14.12 of IPS.

To continue the steps shown in Figure 14.12 of IPS, take 5 random samples of size 50 from the time data and make a histogram of each as shown in the left column of Figure 14.12 in IPS. These have been added to Figure 14.10 on the following page. The means for each sample, along with the population mean, are shown in Table 14.2.

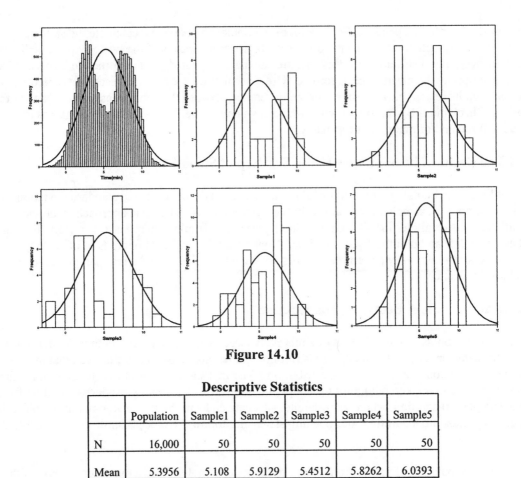

Figure 14.10

Descriptive Statistics

	Population	Sample1	Sample2	Sample3	Sample4	Sample5
N	16,000	50	50	50	50	50
Mean	5.3956	5.108	5.9129	5.4512	5.8262	6.0393

Table 14.2

To continue exploring the question of the accuracy of the bootstrap distribution, adapt the syntax from the bootstrapping commands at the end of the chapter, and then generate a bootstrap distribution for each of the five samples that you just generated. The results from my runs are shown on the following page in Figure 14.11. See Table 14.3 for a summary of the means generated by this process. Yours should look approximately the same (keeping in mind the effects of random variation). Compare these to the distributions shown in the middle column of Figure 14.12 in IPS.

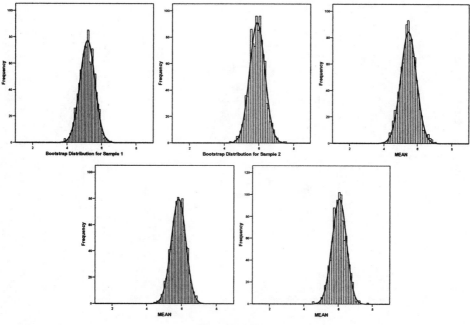

Figure 14.11

Descriptive Statistics

	Population	Bootstrap Distribution for Sample1	Bootstrap Distribution for Sample2	Bootstrap Distribution for Sample3	Bootstrap Distribution for Sample4	Bootstrap Distribution for Sample5
N	16,000	1000	1000	1000	1000	1000
Mean	5.3956	5.9209	6.0951	5.4635	6.0285	6.0515

Table 14.3

Finally, to complete this exercise, we obtain five additional bootstrap distributions for Sample 1. The results from my data set are shown in Figure 14.12 following and the means for these additional bootstrap distributions are shown in Table 14.1. These closely match the right hand column in Figure 14.12 of IPS.

This exercise gives us both numeric and visual information about the bootstrapping distribution. First, each bootstrap distribution has an overall mean that is close to, although not identical to the mean of its original sample. Look at Tables 14.2 and 14.3 earlier. When we compare the mean of Sample 1 (2.5934) with the mean of the bootstrap distribution for Sample 1 (2.5886) we see that they are very close. The same is true for each of our five samples. This can be further established by looking at the means in Table 14.4 for repeated resamples from Sample 1. Thus the bootstrap gives an unbiased estimate for the mean of the original sample.

The shape and spread, as illustrated in Figures 14.10, 14.11, and 14.12, closely resemble the normal sampling distribution. From this we conclude that the shape and spread of the bootstrap distribution depend on the original sample. However, the variation from sample to sample remains.

Finally, look at the six bootstrap distributions for Sample 1. These are included in Figures 14.11 and 14.12. Because these are all so similar, we can conclude that random resampling does not inflate variation.

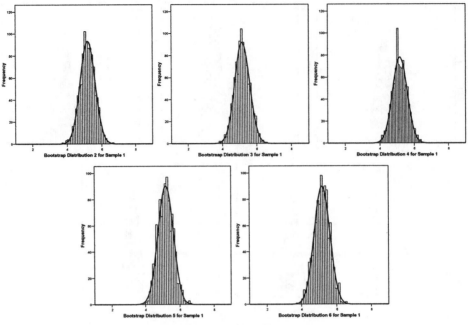

Figure 14.12

Descriptive Statistics

	Population	Bootstrap Distribution 2 for Sample1	Bootstrap Distribution 3 for Sample1	Bootstrap Distribution 4 for Sample1	Bootstrap Distribution 5 for Sample1	Bootstrap Distribution 6 for Sample1
N	16,000	1000	1000	1000	1000	1000
Mean	5.3956	5.1924	5.1705	5.1717	5.1676	5.1695

Table 14.4

Bootstrapping Small Samples

In general, statisticians caution you to use samples as large as possible to reduce error variation. The same is true for bootstrapping. What happens to the bootstrap distribution for small sample sizes? We can address this question by repeating the set of exercises in the previous section with a sample size of 9 instead of 50. To replicate the example shown in IPS Figure 14.13, we start with a normal population. Let us return to the Part Retrieval Times from the previous section. To make our data match the example in Figure 14.13 of IPS, we need a normal distribution. So, I have eliminated the lower half of the original data set used in that example. (Recall that I had created the bimodal distribution by combining two normal distributions.) For our new population, the mean (μ) = 7.9886. Now take 5 samples of size 9 from your population and produce histograms for them. See Figure 14.13 on the following page. These figures match (in general) the figures in the left-hand column of Figure 14.13 in IPS. The means for each of these samples are shown in Table 14.5.

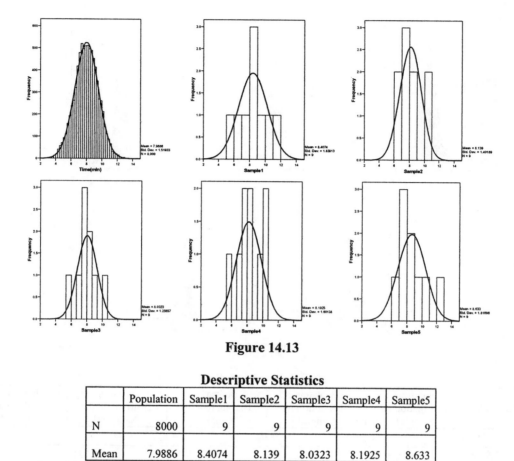

Figure 14.13

Descriptive Statistics

	Population	Sample1	Sample2	Sample3	Sample4	Sample5
N	8000	9	9	9	9	9
Mean	7.9886	8.4074	8.139	8.0323	8.1925	8.633

Table 14.5

Next we get the bootstrap distributions for each of our samples. The results are shown in Figure 14.14 and Table 14.6. Compare these distributions to the ones in the center column of Figure 14.13 in IPS.

Finally, we run repeated resamples of size 9 taken from the first sample. The results are shown in Figure 14.15 and Table 14.7. Compare these to the right-hand column in Figure 14.13 in IPS.

What can we learn from this exercise? Almost all variation in bootstrap distributions comes from the selection of the original sample. The larger the original sample, the smaller this variation will be. Further, small samples are accompanied by problems that bootstrapping cannot overcome. Always be cautious about inferences made from small samples, even when using the bootstrap distribution. Finally, the bootstrap distribution, especially when the resampling process is repeated at least 1000 times, does not introduce appreciable additional variation.

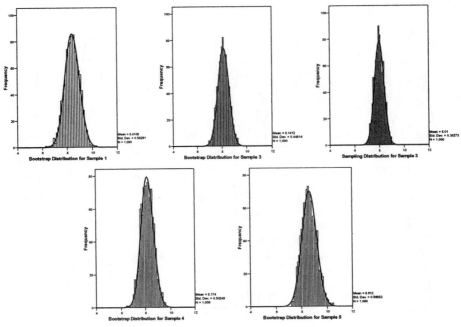

Figure 14.14

Descriptive Statistics

	Population	Bootstrap Distribution for Sample1	Bootstrap Distribution for Sample2	Bootstrap Distribution for Sample3	Bootstrap Distribution for Sample4	Bootstrap Distribution for Sample5
N	8000	9	9	9	9	9
Mean	7.9886	8.4108	8.1412	8.01	8.174	8.612

Table 14.6

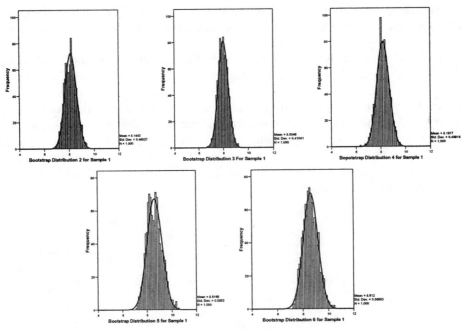

Figure 14.15

Descriptive Statistics

	Population	Bootstrap Distribution 2 for Sample1	Bootstrap Distribution 3 for Sample1	Bootstrap Distribution 4 for Sample1	Bootstrap Distribution 5 for Sample1	Bootstrap Distribution 6 for Sample1
N	8000	9	9	9	9	9
Mean	7.9886	8.1143	8.0046	8.1917	8.6166	8.612

Table 14.7

Bootstrapping A Sample Median

In an earlier section of this chapter, we looked at the 25% trimmed mean as an alternative to using the median. However, we can use the bootstrap procedures for any statistics of interest. To illustrate, we can replicate Figures 14.12 and 14.13 in IPS using the median. For continuity in this manual, I will return to the Part Retrieval time data using the full bimodal data set. The figures and tables that follow are exact parallels to the previous two sections of the manual with the exception that they reference the median of each distribution rather than the mean. The data sets used are the same as those used to produce Figures 14.10 to 14.12 earlier. If you have run the bootstrapping macro you will have noticed that the median is part of the output. If you did not save those files, repeat the runs now and compare your output to the ones in this manual and your text.

Table 14.8 shown below gives the median (50th percentile) value for each of the five samples. The overall median for the population of parts retrieval times is 5.351. Notice that the median for each of the samples is slightly different from the median of the overall data set despite the sample size ($n = 50$). The histogram for the original data set is shown in the upper left of Figure 14.10 earlier. You can superimpose the median value if you wish using the chart options in SPSS.

	Median
Original Sample	5.3510
Sample 1	4.0334
Sample 2	6.7731
Sample 3	6.7367
Sample 4	6.1548
Sample 5	6.1545

Table 14.8

The first step in observing the effects of bootstrapping the median is to bootstrap it for each of the five samples. Histograms of these outcomes are shown in Figure 14.16 on the following page.

Figure 14.17 shows the repeated bootstrap distributions for Sample 1. These distributions are similar to each other indicating that resampling adds little variation. The median values for each of our samples vary considerably among each other and differ from the median value of the original sample (see Table 14.8, above). The bootstrapped distributions for each of the samples also differ from each other as evidenced in Figure 14.16 earlier. Recall that the median value is the one that falls "in the middle" of the distribution. Therefore, the bootstrap distribution repeats the same few values of the data set and these values depend on the original sample. Although standard errors and confidence intervals have not been included in the output shown above, they are reasonably accurate for larger samples ($n = 50$ or greater, for example). Such statistics are less reliable for small samples and for measures such as quartiles that also use only a few numbers in the original sample.

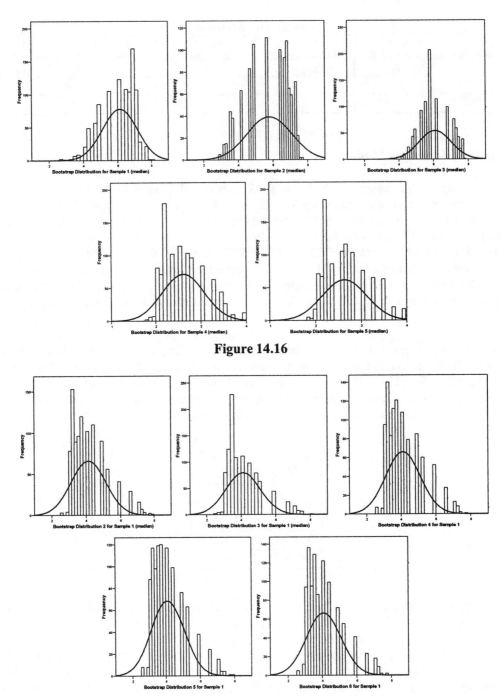

Figure 14.16

Figure 14.17

14.4 Bootstrap Confidence Intervals

Bootstrap Percentile Confidence Intervals

Go back to the SPSS Output Viewer for any of the bootstrap analyses that you have run thus far. You will notice that descriptives for each of the bootstrap distributions (median, mean, and bias) are printed out for you following the computations. Included in the descriptives are the values for the 95% confidence interval for the bootstrap distribution for the statistic. Shown below, in Table 14.9, are the Descriptives for the mean for one of our analyses already completed. The mean that is referenced here is the mean of the bootstrap distribution for the mean of our 1000 resamples. The 95% confidence interval shown is the bootstrap confidence interval.

			Statistic	Std. Error
MEAN	Mean		5.1662	.01356
	95% Confidence Interval for Mean	Lower Bound	5.1396	
		Upper Bound	5.1928	
	5% Trimmed Mean		5.1644	
	Median		5.1707	
	Variance		.1840	
	Std. Deviation		.42889	
	Minimum		3.9100	
	Maximum		6.7500	
	Range		2.8400	
	Interquartile Range		.6300	
	Skewness		.0630	.07700
	Kurtosis		-.1490	.15500

Table 14.9

You may have also noticed that I have included a measure of bias and that descriptive statistics are also printed out for it. Bias is calculated as the difference between the mean of each resample and the mean of the original data set. Small values for the bias measure indicate that the bias in the bootstrap distribution is small. Table 14.10 on the following page shows the descriptive statistics for the bias measure for the same set of data as is included in Table 14.9. Recall that for our original sample we had values between 3 and 15, which is quite a wide spread. In contrast, for all 1000 of our bootstrap replications, the bias measure has a mean of -.2294. In other words, we have, on average, underestimated the mean of the original sample by a very small bit with our multiple resamples. The 95% bootstrap confidence interval for this bias estimate is -.2028 to -.2560. This indicates that the bias of the bootstrap distribution for the mean is small.

Confidence Intervals For Correlation

We can use bootstrapping to obtain the means, standard errors, and confidence intervals for a wide variety of statistics. In this section we will use the bootstrap to obtain the bootstrap distribution for a correlation coefficient. Example 14.9 and Table 14.2 in IPS give us data for a random sample of 50 Major League Baseball players. We want to test the assumption that there is a relationship between performance (measured as the batting average) and salary paid to the players. For the sample, there is a weak correlation of .107 between salary and batting average suggesting that there is only a small relationship between salary and performance.

We can further explore this issue by using the bootstrapping techniques presented earlier to obtain a large number of resamples from this original sample and then checking the 95% confidence interval for them. If the confidence interval covers 0, the observed correlation is not significant.

			Statistic	Std. Error
BIAS	Mean		-.2294	.0136
	95% Confidence Interval for Mean	Lower Bound	-.2560	
		Upper Bound	-.2028	
	5% Trimmed Mean		-.2312	
	Median		-.2249	
	Variance		.1840	
	Std. Deviation		.4289	
	Minimum		-1.4900	
	Maximum		1.3600	
	Range		2.8400	
	Interquartile Range		.6300	
	Skewness		.0630	.0770
	Kurtosis		-.1490	.1550

Table 14.10

There are macros located at www.whfreeman.com/ips5e that will give you the bootstrapping distributions for many of the coefficients produced by the correlation and regression commands in SPSS. Once you have downloaded a macro, make sure that the variable names are correct and that the path and file names are correct for your system. When you run this set of syntax commands, browse the histograms and tables that are printed out looking for information about the 95% confidence interval for the correlation coefficient. Browse the remainder of the output to see, for example, how the significance level fluctuated over the resampling process. Shown in Table 14.11 is the 95% confidence interval for several of the relevant variables. Check to see if the 95% confidence interval for the correlation coefficient covers 0 or not. For reference, the overall significance level is shown in this table as well. Note that the confidence interval for the correlation coefficient and for the significance both give you the same answer about whether or not the observed overall correlation is significant. Figure 14.18 shows the bootstrap distribution of the correlation coefficients. Shown to the right of the histogram is the mean and standard deviation of all of the correlation coefficients produced. Compare this figure to Figure 14.16 in IPS. The mean of the correlation coefficients for all 1000 samples is .10968 which compares with .107 for the original sample.

Figure 14.18

Statistics

		Salary_BatAvg_PearsonCorrelation	Salary_BatAvg_Sig.2tailed
N	Valid	1000	1000
	Missing	0	0
Percentiles	2.5	-.14916	.00930
	97.5	.36430	.96604

Table 14.11

Overall, these data do not give us evidence of a significant relationship between salary and batting average.

14.5 Significance-Testing Using Permutation Tests

Significance tests tell us whether or not an observed effect could reasonably occur "by chance." The reasoning for hypothesis tests is reviewed in IPS. The null hypothesis makes the statement that the effect we seek is not present in the population. Small p values are evidence against the null hypothesis and in favor of a real effect in the population. Tests based on resampling use the same reasoning. In resampling the p value is based on calculations rather than formulas and as a result, they can be used in situations in which the traditional tests are not applicable.

Permutation Tests For Two Independent Groups

Resampling for significance tests we estimate the sampling distribution of the test statistic when the null hypothesis is true. That is, we resample as if the null hypothesis is true. Example 14.11 in IPS revisits the Degree of Reading Power (DRP scores) from earlier analyses. To apply the idea of a permutation test to these data, we start with the difference between the sample means as a measure of the effect of "directed reading activities." If the null hypothesis is true, then the distribution of DRP scores is independent of teaching method of the group that the children are assigned to.

To resample in a way that is consistent with the null hypothesis, we repeatedly take a random selection of student scores and then assign them to treatment groups. That is, we sample without replacement in permutation tests. Therefore this sampling procedure is called a permutation resample to differentiate it from the resamples we met earlier in this chapter.

To complete the permutation tests for Example 14.11, go to the Web site at www.whfreeman.com/ips5e and download the macro called "Bootstrapping for Permutation Tests Indep T-Tests" and run it. It is set up to work with the data in Table 14.3. Set the path to this file to match your system. This bootstrapping macro first randomly selects 21 of the 44 scores (students) to be in the treatment group. The other 23 student scores are assigned to the control group. This is consistent with a null hypothesis that states that there is no difference between the groups. Once the student scores are randomly assigned to groups, the macro will calculate the difference between the mean of the two groups. This resampling process is then repeated 1000 times. The difference between the means is the statistic of interest and the macro will give you a histogram with a normal curve superimposed and the mean and standard deviation of this statistic.

Figure 14.19 gives us the distribution of our difference between the means when the null hypothesis is true. Compare this to Figure 14.21 in IPS. Notice that this distribution is centered at zero (no effect) as the null hypothesis predicts. Superimposed over this figure is a vertical line representing the difference between the means of the original samples. Differences greater than this value give us evidence in favor of the alternative hypothesis. To calculate the p value associated with the outcome of these permutation tests, the permutation test p value, we find the proportion of the 1000 samples that had a difference greater than or equal to 9.954. A review of the data produced by this run of the macro gives us 9 mean differences greater than the cutoff point. To calculate the permutation test p value, we add +1 to both pour numerator and denominator to get $(9 + 1)/(1000 + 1) = .00999$ or $.010$.

Figure 14.19 below shows that the permutation distribution has a roughly normal shape. This distribution approximates the sampling distribution, therefore, we now know that the sampling distribution is close to normal so we can apply the usual two-sample t test. The permutation p value calculated above is close to $.013$ which we calculated as the p value for this independent t test back in Chapter 7 (Example 7.14).

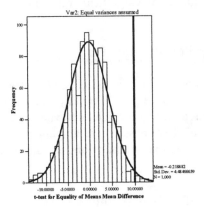

Figure 14.19

Permutation Tests For Matched Pairs

We have now established the general procedure for permutation tests. There are three steps:

1. Compute the statistic for the original data.
2. Choose permutation resamples from the data without replacement. Select the data in accordance with the null hypothesis. Now construct the permutation distribution of the statistic form a large number of resamples.
3. Find the p value by locating the original statistic on the permutation distribution.

We can repeat this procedure for a matched pairs t test. Go to the Web site at www.whfreeman.com/ips5e and download the macro called "Bootstrapping for Permutation Tests Matched Pairs T-Tests". This macro is set up to use the data from Table 14.4 for the "moon days" study of dementia. Be certain that the path to the data is set for your system.

The null hypothesis is that there is no difference between "moon" and "other" days. Resampling in accordance with this hypothesis randomly assigns one of each patient's two scores to "moon" and one to "other." We must maintain the original pairing of scores but randomly assign them to condition. Figure 14.20 shows the permutation distribution of the matched pairs differences for 1000 resamples. It is close to a normal distribution. Superimposed on the figure is the value for the observed difference of 2.433. This value has been superimposed on Figure 14.20. Notice that none of these resamples produces a difference as large as the observed difference. The estimated p value is therefore $(0 + 1)/(1000 + 1) = .001$. Although we have fewer resamples and therefore a larger p value than in the text example, we still have strong evidence that aggressive behavior is more common on "moon" days.

Given the near normality of the permutation distribution, we conclude that the paired t test is a reliable test for these data.

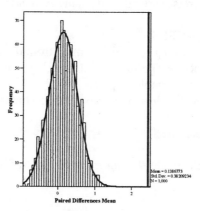

Figure 14.20

Permutation Tests For The Significance Of A Relationship

Example 14.9 and Table 14.2 contain data for the relationship between the batting average and salary of 50 randomly selected baseball players. The null hypothesis for the correlation between these two variables is that $r = 0$. We can test this hypothesis using permutation tests as we have for matched and independent groups t tests. To replicate the analysis shown here, go to the text Web site and download a macro called "Bootstrapping for Permutation Tests Correlation.' In this example we have maintained the order of the batting averages and shuffled the order of the salaries, then calculated the correlation between the variables for each of 1000 resamples. The p value is calculated as in other examples. Using the previous two examples as models, calculate the p value and write the summary of these findings.

Exercises For Chapter 14

Note: The data sets used in these exercises are taken from the Appendix of your text.

1. Open the data set called BIOMARKERS.
    a. Write your own syntax for sampling with replacement from the variable called *trap.*
    b. Using the macro called "BootstrapTheMeanandMedian," generate a bootstrap sampling distribution for the mean of the variable *trap.*
    i. Be sure to change the mean of the distribution to the mean for *trap* (in the original sample) so that the bias measure is correct.
    c. Write a description of the outcome.

2. Repeat Question 1 above for the variable *flaking* from the dandruff data set.

3. Open the data set called PNG.
    a. Graph the distribution of *retinol.*
    b. Bootstrap the median for this variable.
    c. Create the bootstrap distribution for the median.
    d. How does the shape of the bootstrap distribution compare to the original shape of the histogram for this variable?
    e. What is the bootstrap standard error for the median?
    f. Calculate the 95% confidence interval for the median.
    g. Write a brief summary of the outcome for *retinol.*

4. Open the data set called BIOMASS.
    a. Bootstrap the correlation coefficient for the relationship between *voplus* and *vominus.*
    b. Graph the bootstrap distribution for the correlations.
    c. Calculate the 95% confidence intervals for the bootstrap distribution.
    d. Do these data give us evidence of a significant relationship between *voplus* and *vominus?*

5. Open the data set called csdata.
    a. State the null hypothesis in terms of permutation tests for the difference between men and women (*sex*) for *satm* scores
    b. Complete permutation tests to test the significance of the difference between men and women (*sex*) for *satm* scores.
    c. What does the sampling distribution for the difference between the means for each sex tell us? What shape does it take? Where is it centered?
    d. Superimpose a line on the graph to represent the difference between men and women from the original data set.
    e. Calculate the (most accurate) *p* value from these data.
    f. Can we appropriately apply the two sample *t* test?
    g. Summarize your outcome.

6. Open the data set called READING.
    a. State the general steps for permutation tests.
    b. Bootstrap the permutation tests for matched pairs for the variables *pre2* and *post3.*
    c. Find the *p* value for the *t* test for matched pairs from the permutation tests.

7. Using the data set described in Question 6, complete permutation tests for the significance of the relationship between the variables *pre2* and *post3.*

Chapter 15. Nonparametric Tests

Topics covered in this chapter:

15.1 Wilcoxon Rank Sum Test
15.2 Wilcoxon Signed Rank Test
    Ties
15.3 Kruskal-Wallis Test

This chapter introduces one class of **nonparametric** procedures, tests that can replace *t* tests and one-way analysis of variance when the normality assumption for those tests is not met. When distributions are strongly skewed, the mean may not be the preferred measure of center. The focus of these tests is on medians rather than means. All three of these tests use ranks of the observations in calculating the test statistic.

15.1 Wilcoxon Rank Sum Test

The **Wilcoxon rank sum test** is the nonparametric counterpart of the parametric independent *t* test. It is applied to situations in which the normality assumption underlying the parametric independent *t* test has been violated or questionably met. The focus of this test is on medians rather than means. An alternate form of this test, the one used by SPSS for Windows, is the Mann-Whitney *U* test.

Example 15.1 in IPS gives us the following example. Small numbers of weeds may reduce the corn yields per acre of farmland. In eight small plots of land, a researcher planted corn at the same rate in each and then weeded the corn rows by hand to make certain that no weeds were allowed to grow in four of these plots (randomly selected). Lamb's-quarter, a common weed in corn fields, was allowed to grow in the remaining four plots at the rate of three lamb's-quarter plants per meter of row.

Table 15.1 presents the yields of corn for each of these two types of plots.

Weeds Per Meter	Yield (bu/acre)			
0	166.67	172.2	165.0	176.9
3	158.6	176.4	153.1	156.0

Table 15.1

We would like to know whether lamb's-quarter at this rate reduces yield. Whereas a Chi-square test could be applied to answer the general question, this test ignores the ordering of the responses and so does not use all of the available information. Because the data are ordinal, a test based on ranks makes sense. One can use the Wilcoxon rank sum test for the hypotheses:

    H_0: the two types of plots do not differ in their yields
    H_a: one of the two plots gives systematically larger yields than the other.

The data were entered into SPSS for Windows using 8 rows and 2 columns with the variable names Weeds (declared numeric with value labels 1 = No Weeds and 2 = 3 Weeds), and Yields.
It is important to note that the grouping variable (in this case, Weeds) must be a numeric variable, not a string variable.

To conduct a Wilcoxon rank sum test (or Mann-Whitney *U* test), follow these steps:

1. Click **Analyze,** click **Nonparametric Tests,** and then click **2 Independent Samples.** The "Two-Independent-Samples Tests" window in Figure 15.1 appears.

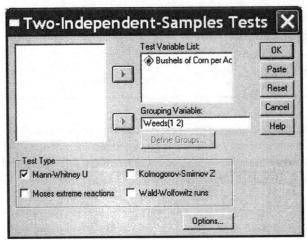

Figure 15.1

2. Click *yield,* then click ▸ to move *yield* into the "Test Variable List" box.
3. Click *weeds,* then click ▸ to move *weeds* into the "Grouping Variable" box.
4. Click the **Define Groups** button.
5. Type *1* in the "Group 1" box, press the **Tab** key, and type *2* in the "Group 2" box (to match the defined levels of your variable).
6. Click **Continue.**
7. The default test is the Mann-Whitney *U* test (as indicated by the ✔ in front of "Mann-Whitney U" in the "Test Type" area).
8. Click **OK.**

As can be seen in Table 15.2, the rank sum for 0 Weeds per Meter is $W = 23$.

Ranks

	Lamb's-quarter per meter	N	Mean Rank	Sum of Ranks
Bushels of Corn per Acre	0 Weeds per Meter	4	5.75	23.00
	3 Weeds per Meter	4	3.25	13.00
	Total	8		

Table 15.2

The Mann-Whitney U and Wilcoxon W test statistics are shown in Table 15.3. Note that the Z level matches that shown in IPS Example 15.4. The *z* distribution should yield a reasonable approximation of the *p* value. As shown in Table 15.3, the standardized value is $z = -1.443$ (negative because we subtracted in the opposite direction as the example in IPS) with a two-sided *p* value = .200 (Example 15.4 of IPS lists the *p* value as 0.10 —which is .200 divided by 2). This moderately large *p* value provides little evidence that yields are reduced by small numbers of weeds.

Test Statistics(b)

	Bushels of Corn per Acre
Mann-Whitney U	3.000
Wilcoxon W	13.000
Z	-1.443
Asymp. Sig. (2-tailed)	.149
Exact Sig. [2*(1-tailed	.200(a)

a Not corrected for ties.
b Grouping Variable: Lamb's-quarter per meter
Table 15.3

SPSS shows the significance levels for two-tailed tests.

Example 15.6 in IPS gives us a demonstration of the use of null and alternate hypotheses for Wilcoxon tests. A brief summary follows. Food sold at outdoor fairs and festivals may be less safe than food sold in restaurants because it is prepared in temporary locations and often by volunteer help. What do people who attend fairs think about the safety of food served? One study asked this question of people at a number of fairs in the Midwest:

How often do you think people become sick because of food they consume prepared at outdoor fairs and festivals? The possible responses were: 1 = very rarely, 2 = once in a while, 3 = often, 4 = more often than not, and 5 = always.

In all, 303 people answered the question. Of these, 196 were women and 107 were men. Is there good evidence that men and women differ in their perceptions about food safety at fairs? The data can be retrieved from the Web site: www.whfreeman.com/ips5e.

We would like to know whether men or women are more concerned about food safety. Whereas a Chi-square test could be applied to answer the general question, it ignores the ordering of the responses and so does not use all of the available information. Because the data are ordinal, a test based on ranks makes sense. One can use the Wilcoxon rank sum test for the hypotheses:

H_0: men and women do not differ in their responses
H_a: one of the two genders gives systematically larger responses than the other.

The data were entered into the SPSS Data Editor using 10 rows and 3 columns with the variable names gender (declared numeric 8.0 with value labels 1 = Female and 2 = Male), sick (declared numeric 8.0 with value labels 1 = very rarely, 2 = once in a while, 3 = often, 4 = more often than not, and 5 = always), and weight (declared numeric 8.0), where weight represents the count of individuals for each gender who selected each of the five response options.

It is important to note that the grouping variable (in this case, gender) must be a numeric variable (not entered as M's and F's). Also, before performing analyses, the weighting option under **Data** and **Weight Cases** was activated. See Figure 15.2.

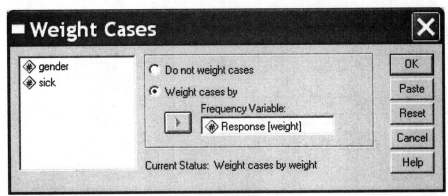

Figure 15.2

To conduct a Wilcoxon rank sum test (or Mann-Whitney U test), follow the steps shown above for the corn and lamb's-quarter example.

The "Two-Independent-Samples Tests" window in Figure 15.3 appears.

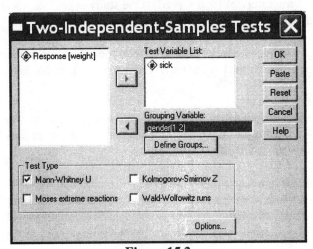

Figure 15.3

The resulting output is displayed in Tables 15.4, below, and 15.5 on the following page. As can be seen in Table 15.4, the rank sum for men (using average ranks for ties) is $W = 14,059.5$. Because the sample size is large, the z distribution should yield a reasonable approximation of the p value.

Ranks

	gender	N	Mean Rank	Sum of Ranks
sick	Female	196	163.25	31996.50
	Male	107	131.40	14059.50
	Total	303		

Table 15.4

As shown in Table 15.5 on the following page, the standardized value is $z = -3.334$ with a two-sided p value $= 0.001$. This small p value lends strong evidence that women are more concerned than men about the safety of food served at fairs. These results match those shown in IPS Example 15.7 and IPS Figure 15.5.

Test Statistics(a)

	sick
Mann-Whitney U	8281.500
Wilcoxon W	14059.500
Z	-3.334
Asymp. Sig. (2-tailed)	.001

a Grouping Variable: gender

Table 15.5

15.2 Wilcoxon Signed Rank Test

This section will introduce the **Wilcoxon signed rank test,** the nonparametric counterpart of a paired-samples *t* test. It is used in situations in which there are repeated measures (the same group is assessed on the same measure on two occasions) or matched subjects (pairs of individuals are each assessed once on a measure). It is applied to situations in which the assumptions underlying the parametric *t* test have been violated or questionably met. The focus of this test is on medians rather than means.

Example 15.8 in IPS gives us data for story telling by children and uses only the low-progress readers. The data for this study can be found at www.whfreeman.com/ips5e. First we complete a t-test for the data. The results are shown below in Table 15.6.

Paired Samples Test

		Paired Differences					t	df	Sig. (2-tailed)
		Mean	Std. Deviation	Std. Error Mean	95% Confidence Interval of the Difference				
					Lower	Upper			
Pair 1	Story 1 Score - Story 2 Score	-.11000	.38736	.17323	-.59097	.37097	-.635	4	.560

Table 15.6

To complete the Wilcoxon Signed Rank Test, follow the instructions given earlier in the chapter for the Wilcoxon test except instead of choosing **2 Independent Samples,** under **Nonparametric Tests** choose **2 Related Samples** and the screen shown in Figure 15.4 on the facing page will appear. Click on Score 1 and Score 2 and then move them into the box labeled "Test Pair(s) List". Click on Wilcoxon if there is not already a check mark beside this choice in the "Test Type" box. Click **OK.**

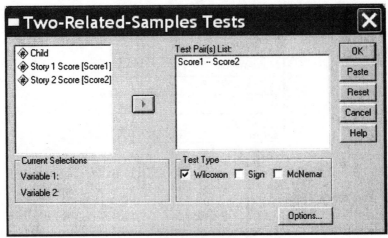

Figure 15.4

The results are shown in Tables 15.7 and 15.8. SPSS calculated W for Positive Ranks (W⁺) = 9.0 as in IPS. *Z* is equal to -.405 and *p* = .686. These results indicate a nonsignificant difference in the story telling scores with and without illustration. This is the same outcome as for the t-test presented in Table 15.6 earlier.

Ranks

		N	Mean Rank	Sum of Ranks
Story 2 Score - Story 1 Score	Negative Ranks	3(a)	2.00	6.00
	Positive Ranks	2(b)	4.50	9.00
	Ties	0(c)		
	Total	5		

a Story 2 Score < Story 1 Score
b Story 2 Score > Story 1 Score
c Story 2 Score = Story 1 Score

Table 15.7

Test Statistics(b)

	Story 2 Score – Story 1 Score
Z	-.405(a)
Asymp. Sig. (2-tailed)	.686

a Based on negative ranks.
b Wilcoxon Signed Ranks Test

Table 15.8

Ties

Example 15.11 in IPS gives us the golf scores of 12 members of a college women's golf team in two rounds of tournament play. A golf score is the number of strokes required to complete the course, therefore, low scores are better. The data were entered into SPSS for Windows using two columns and the variable names **round_1** (declared numeric 8.0) and **round_2** (declared numeric 8.0). The variables *round_1* and *round_2* were entered into the first and second columns, respectively. These data can be retrieved from www.whfreeman.com/ips5e/.

Because this is a matched pairs design, inference is based on the differences between pairs. Negative differences indicate better (lower) scores on the second round. We see that 6 of the 12 golfers improved their scores. We would like to test the hypotheses that in a large population of collegiate women golfers:

H_0: scores have the same distribution in Rounds 1 and 2
H_a: scores are systematically lower or higher in Round 2.

The assessment of whether the assumption of normality has been met is based on the difference in golf scores. A small sample makes it difficult to assess normality adequately, but the normal quantile plot of the differences in Figure 15.5 shows some irregularity and a low outlier.

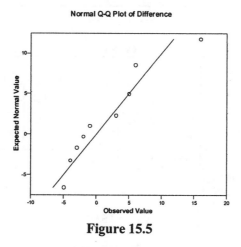

Figure 15.5

To conduct a Wilcoxon signed rank test, follow these steps:

1. Click **Analyze,** click **Nonparametric Tests,** and then click **2 Related Samples.** The "Two-Related-Samples Tests" window shown in Figure 15.6, on the facing page, appears.
2. Click *round_1* and it appears after "Variable 1" in the "Current Selections" box.
3. Click *round_2* and it appears after "Variable 2" in the "Current Selections" box.
4. Click ‣ to move the variables into the "Test Pair(s) List" box (it will read "*round_1 – round_2*").
5. The default test is the Wilcoxon signed rank test (as indicated by the ✔ in front of Wilcoxon in the "Test Type" box).
6. Click **OK.**

The resulting SPSS for Windows output is displayed in Tables 15.9 and 15.10 on the facing page. First, notice that the difference in Table 15.9 reads "round_2 - round_1." For this test, SPSS for Windows always creates a difference score between the two named variables based on the order in which the variables are entered in the dataset. The variable that appears *first* in the dataset is always subtracted from the variable that appears *later* in the dataset. For this problem, the variables were entered in the order of *round_1* and then *round_2.* Thus, SPSS for Windows creates a difference score of *round_2* minus *round_1,* despite the order in which the variables appear in the "Test(s) Pairs List" box in the "Two-Related-Samples Tests" window. The same conclusion will be reached regardless of the order in which subtraction was done because the two-tailed *p* value will be the same whether the difference is *round_2 – round_1* or *round_1 – round_2.* However, caution must be used in performing a directional test.

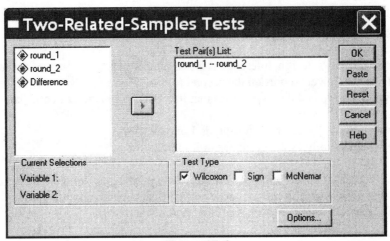

Figure 15.6

Ranks

		N	Mean Rank	Sum of Ranks
round_2 - round_1	Negative Ranks	6(a)	4.58	27.50
	Positive Ranks	6(b)	8.42	50.50
	Ties	0(c)		
	Total	12		

a round_2 < round_1
b round_2 > round_1
c round_2 = round_1

Table 15.9

Test Statistics(b)

	round_2 - round_1
Z	-.910(a)
Asymp. Sig. (2-tailed)	.363

a Based on negative ranks.
b Wilcoxon Signed Ranks Test

Table 15.10

As shown in Table 15.9, the sum of the Negative Ranks (round_2 - round_1) is 27.5 and the sum of the Positive Ranks (round_2 > round_1) is W^+=50.5. The value of 0 for Ties means that there were no pairs of scores in which the values were the same (e.g., round_2 = round_1 [it does not mean that there were no ties among the ranks]). The Wilcoxon signed rank statistic is the sum of the positive differences. In Table 15.10, a z value of -0.91 is reported that is based on the standardized sum of the positive ranks and is not adjusted for the continuity correction. The corresponding p value is given as 0.363. These data give weak evidence for a systematic change in scores between rounds.

15.3 Kruskal-Wallis Test

The **Kruskal-Wallis test** is the nonparametric counterpart of the parametric one-way analysis of variance. It is applied to situations in which the normality assumption underlying the parametric one-way ANOVA has been violated or questionably met. The focus of this test is on medians rather than means.

Example 15.13 in IPS tells us that lamb's-quarter is a common weed that interferes with the growth of corn. A researcher planted corn at the same rate in 16 small plots of ground and then randomly assigned plots to four groups. He weeded the plots by hand to allow a fixed number of Lamb's-quarters to grow in each meter of a corn row. These numbers were 0, 1, 3, and 9 in the four groups of plots. No other weeds were allowed to grow, and all plots received identical treatment except for the weeds. The summary statistics for the data are shown in Table 15.11.

Weeds per meter	n	Mean	Std. Dev.
0	4	170.200	5.422
1	4	162.825	4.469
3	4	161.025	10.493
9	4	157.575	10.118

Table 15.11

The sample standard deviations do not satisfy the rule of thumb from IPS for use of ANOVA that the largest standard deviation should not exceed twice the smallest. Moreover, we see that outliers are present in the yields for 3 and 9 weeds per meter. These are the correct yields for their plots, so we have no justification for removing them. We may want to use a nonparametric test.

The hypotheses are:

H_0: yields have the same distribution in all groups
H_a: yields are systematically higher in some groups than in others.

The data were entered in two columns using the variables *weeds* (declared numeric 8.0) and *yield* (declared numeric 8.1). It is important to note that the grouping variable (in this case, *weeds*) must always be a numeric variable. To conduct a Kruskal-Wallis *H* test, follow these steps.

1. Click **Analyze,** click **Nonparametric Tests,** and then click **K Independent Samples.** The window shown in Figure 15.7 will appear.
2. Click *yield,* then click ▸ to move *yield* into the "Test Variable List" box.
3. Click *weeds,* then click ▸ to move *weeds* into the "Grouping Variable" box.
4. Click **Define Range.**
5. Because 0 to 9 is the range for this example, type *0* in the "Minimum" box, press the **Tab** key, and type *9* in the "Maximum" box.
6. Click **Continue.**
7. The default test is the Kruskal-Wallis H test (as indicated by the ✔ in front of Kruskal-Wallis H in the "Test Type" box).
8. Click **OK.**

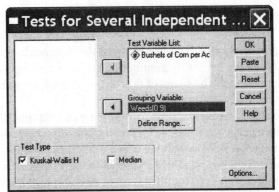

Figure 15.7

Examination of Table 15.11, presented earlier, suggests that an increase in weeds results in decreased yield. As can be seen in Table 15.12, the mean rank for the group with 0 weeds per meter was 13.13, the mean rank for the group with 1 weed per meter was 8.38, the mean rank for the group with 3 weeds per meter was 6.25, and the mean rank for the group with 9 weeds per meter was 6.25. SPSS for Windows uses the Chi-square approximation to obtain a p value = 0.134, as shown in Table 15.13.

Ranks

	Lamb's-quarter per meter	N	Mean Rank
Bushels of Corn per Acre	0 Weeds per meter	4	13.13
	1 Weeds per meter	4	8.38
	3 Weeds per meter	4	6.25
	9 Weeds per meter	4	6.25
	Total	16	

Table 15.12

Test Statistics(a,b)

	Bushels of Corn per Acre
Chi-Square	5.573
df	3
Asymp. Sig.	.134

a Kruskal Wallis Test
b Grouping Variable: Lamb's-quarter per meter

Table 15.13

The Chi-square value calculated by SPSS (5.573) closely matches that shown in Example 15.13 in IPS (5.56). This small experiment suggests that more weeds decrease yield but does not provide convincing evidence that weeds have an effect.

Exercises For Chapter 15

1. Meerkats (*Suricata suricatta*) are social animals native to the dry grassland areas of Africa. They can be seen in most North American zoos and are a delight to watch. As is the case for all carnivores, meerkats are smelly creatures. They scent mark their environment with an anal scent gland. Here are representative data for the number of times each of the members of a captive colony sniffed at such scent marks.

Male	Female
100	48
175	50
83	40
166	55
110	35
180	42

    a. Do a thorough exploratory data analysis including a normal quantile plot.
    b. Choose the appropriate statistical test and write a sentence or two explaining your choice.
    c. Write the appropriate hypotheses for a comparison of the three groups.
    d. Complete the analysis and, using a two-sided test, interpret the outcome for hypothesis testing.

2. The following data approximate the number of inferences children make in their story summaries.

Younger Children	Older Children
0	4
1	4
0	7
0	4
3	6
2	4
3	6
2	8
5	7
2	8

    a. Do a thorough exploratory data analysis including a normal quantile plot.
    b. Choose the appropriate statistical test and write a sentence or two explaining your choice.
    c. Write the appropriate hypotheses for a comparison of the three groups.
    d. Complete the analysis and, using a two-sided test, interpret the outcome for hypothesis testing.

3. Behavioral scientists are interested in the analysis of brain functions. The amygdala is thought to mediate fear responses. In a test of this supposition, a group of rats were given lesions in the amygdala and then their fear responses were compared to those of a control group without lesions. Data are shown below.

Lesions	Control
15	9
14	4
15	9
8	10
7	6
22	6
36	4
19	5
14	9
18	

	17		

a. Do a thorough exploratory data analysis.
b. Choose the appropriate statistical test and write a sentence or two explaining your choice.
c. Write the appropriate hypotheses for a comparison of the three groups.
d. Complete the analysis and, using a two-sided test, interpret the outcome for hypothesis testing.

4. Clinical psychologists are interested in the effectiveness of treatment approaches. In some types of psychiatric disorder, patients have difficulty forming hypotheses. What follows are data for hypothesis formation before and after a treatment designed to improve the performance of this type of psychiatric patient.

Patient	Before	After
1	8	7
2	4	9
3	2	3
4	2	6
5	4	3
6	8	10
7	3	6
8	1	7
9	3	8
10	9	7

a. Do a thorough exploratory data analysis.
b. Choose the appropriate statistical test and write a sentence or two explaining your choice.
c. Write the appropriate hypotheses for a comparison of the three groups.
d. Complete the analysis and, using a two-sided test, interpret the outcome for hypothesis testing.

5. The following data are for three independent groups.

Group 1	Group 2	Group 3
6	8	12
10	12	13
12	9	20
16	15	18
14	17	22
7	11	15

a. Do a thorough exploratory data analysis.
b. Choose the appropriate statistical test and write a sentence or two explaining your choice.
c. Write the appropriate hypotheses for a comparison of the three groups.
d. Complete the analysis and, using a two-sided test, interpret the outcome for hypothesis testing.

6. A researcher investigates the effects of three different levels of a new drug known to improve memory. A test for memory is conducted exactly 1 hour after administration of the drug. Data are presented on the following page. The higher the score, the better the memory.

Drug Level		
1	2	3
6	14	9
7	25	7
10	13	7
8	15	9
12	19	13

a. Do a thorough exploratory data analysis.
b. Choose the appropriate statistical test and write a sentence or two explaining your choice.
c. Write the appropriate hypotheses for a comparison of the three groups.
d. Complete the analysis and, using a two-sided test, interpret the outcome for hypothesis testing.

Chapter 16. Logistic Regression

Topics covered in this chapter:

Logistic regression is used when we have a response variable that has only two values. We also have an explanatory variable as in earlier regression approaches that we explored. In Chapters 5 and 8 we looked at binomial probabilities and odds ratios. We will use these techniques again in this chapter. Since the response variable has only two possible outcomes, we can look at the relationship between the explanatory variable and the odds ratio for the binomial response variable.

When we modeled linear and multiple regression in earlier chapters, we used the equation for a linear relationship between the mean of the response variable as a function of the explanatory variable. The standard formula was stated as $\mu = \beta_0 + \beta_1 x$. For logistic regression, we use a similar formula and the natural logarithms of the odds. That is, we use log $(p/1\text{-}p)$.

16.1 The Logistic Regression Model

We use binary logistic regression to model the event probability for a categorical response variable with two outcomes. For example, a catalog company can send mailings to the people who are most likely to respond; a doctor can determine whether a tumor is more likely to be benign or malignant; and a loan officer can assess the risk of extending credit to a particular customer.   In these cases we are interested in probabilities and the probability of an event must lie between 0 and 1. With linear regression techniques, the linear regression model allows the dependent variable to take values greater than 1 or less than 0. The logistic regression model is a type of generalized linear model that extends the linear regression model by linking the range of real numbers to the 0–1 range.

We will use the CHEESE data taken from the appendix of the text and referenced in Examples 16.5, 16.9, and 16.10. To begin we need to define our *TasteOK* variable. The steps are given here:

1. Click **Transform, Compute.**
2. Enter *TasteOK* in the "Target Variable" box.
3. Enter "0" in the "Numeric Expression" box. This makes all values of *TasteOK* = 0.
4. Click **OK.**
5. Click **Transform, Compute.**
6. Leave *TasteOK* in the "Target Variable" box.
7. Enter "1" in the "Numeric Expression" box.
8. Click **If.**
9. Click beside "Include if case satisfies condition" and enter "Taste GE 37" in the box.

10. Click **Continue.** Click **OK.**
11. SPSS will prompt you with the question "Change existing variable?"
12. Click **OK.**
13. Go to the "Variable View" tab and add labels to the *TasteOK* variable using 1 = "Acceptable" and 0 = "Unacceptable."

Running The Analysis

To create the logistic regression model:

1. Click **Analyze, Regression, Binary Logistic.** See Figure 16.1.
2. Select *TasteOK* as the dependent variable. See Figure 16.2.
3. Select *Acetic* as the covariate.
4. Click **Continue.**
5. Click "Options" in the "Logistic Regression" dialog box.
6. Select 95% "CI for Exp(B) ." See Figure 16.3.
7. Click **Continue.**
8. Click **OK** in the Logistic Regression dialog box.

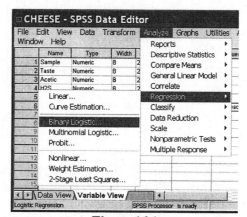

Figure 16.1

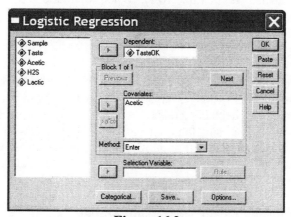

Figure 16.2

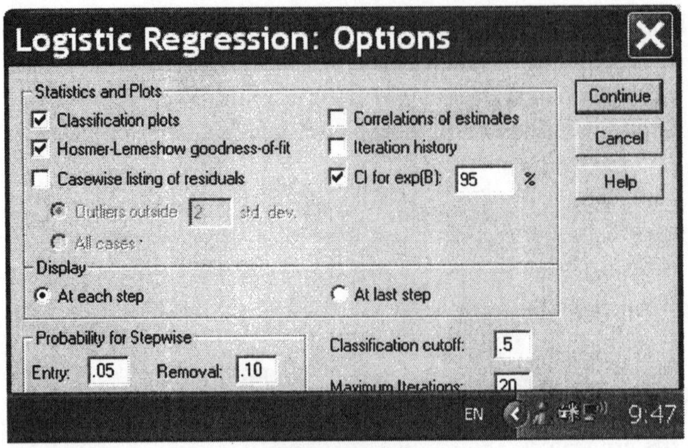

Figure 16.3

Look for the "Variables in the Equation" box on the output file as shown in Table 16.1 below.

Variables in the Equation

		B	S.E.	Wald	df	Sig.	Exp(B)	95.0% C.I. for EXP(B)	
								Lower	Upper
Step 1(a)	Acetic	2.249	1.027	4.795	1	.029	9.479	1.266	70.959
	Constant	-13.705	5.932	5.338	1	.021	.000		

a Variable(s) entered on step 1: Acetic.

Table 16.1

Binomial Distribution And Odds

Since logistic regression works with odds or the ratio between two proportions, we need to look for the appropriate proportions in the output from SPSS. For this example, our indicator variable is ***TasteOK*** with Acceptable and Unacceptable as the levels of this variable. The odds for "Acceptable" are the usual $p/(1-p)$. The same is true for "Unacceptable." The odds ratio can be calculated using the beta coefficients (in the column labeled "B" in Table 16.1). When we divide -13.705 by 2.249 we get 9.479 as shown in the column labeled "Exp(B)" in Table 16.1.

Model For Logistic Regression

In logistic regression we are interested in the mean of a response variable p in terms of a binomial explanatory variable x. Although not entirely straightforward to do, we can transform the odds value to its natural logarithm and thereby model the log odds as a function of the explanatory variable. Thus we arrive at the logistic regression model using $\log(p/(1-p)) = \beta_0 + \beta_1 x$. In this model, p is a binomial proportion and x is the explanatory variable. For our CHEESE example, these values are 2.249 for Acetic and -3.705 for the Constant and are shown in Table 16.1 above.   The model can be written as:

$$\text{Log(odds)} = b_0 + b_1 x = -13.705 + 2.249\, x$$

Fitting And Interpreting The Logistic Regression Model

Now that we have the model for estimated log odds, we can calculate the odds ratio. SPSS prints it out for you under the heading "Exp(b)." When we look at Table 16.1 above, we find the value 9.479. This tells us that if we increase the Acetic Acid content x by one unit, we increase the odds that the cheese will be acceptable by about 9.5 times.

16.2   Inference For The Logistic Regression Model

Statistical inference for logistic regression is essentially the same as what we did for simple linear regression. SPSS will calculate and print for us the estimates of the model parameters and their standard errors. As well, we will get 95% confidence intervals for these estimates (see Steps 5 and 6 in "Running the Analysis" above). The ratio of the estimate to the standard error is the basis for hypothesis testing. SPSS gives the test statistic as a chi-square (the square of the z statistic). Thus p values are obtained from the chi-square distribution with 1 degree of freedom.

We will continue in this section with the CHEESE dataset to illustrate the SPSS outputs. First, look at the standard error for the slope of β_1. Our table of "Variables in the Equation" shown in Table 16.1 earlier gives us 1.027. The 95% confidence interval for β_1 is calculated as $b_1 \pm z^*\text{SE } b_1$ or $2.249 \pm (1.96)(1.027)$. The 95% confidence interval for the odds ratio is shown to the right of the odds ratio in Table 16.1 and the range is 1.266 to 70.959. To test the hypothesis that $\beta_1 = 0$, we can use the Wald chi square shown in the output as 4.795, with df = 1 and $p = .029$. These values lead us to reject our null hypothesis and match those presented in the text for Example 16.9.

These results lead us to statements like the following. Increasing the acetic acid content of cheese will increase the odds that the cheese is acceptable by about 9.5 times. The odds could be as small as 1.27 or as large as 70.96. Thus we have evidence that cheeses with higher acetic acid content are more likely to be acceptable but we require more data to predict the relationship accurately.

Multiple Logistic Regression

The dataset for the cheese example contains two additional explanatory variables, H2S and Lactic. We can use multiple logistic regression to answer the question: Do these other explanatory variables contain information that will give us a better prediction?

Running The Analysis

To create the multiple logistic regression model using SPSS, follow the steps shown below.

1. Click **Analyze, Regression, Binary Logistic.**
2. Select *TasteOK* as the dependent variable.
3. Select *Acetic, H2S,* and *Lactic* as the covariates.
4. Click "Options" in the "Logistic Regression" dialog box.
5. Select 95% "CI for Exp(B)."
6. Click **Continue.**
7. Click **OK** in the Logistic Regression dialog box.

The output is shown in Table 16.2 following and matches the output shown in Figure 16.7 in IPS. From this information we can construct our prediction equation as:

$$-14.26 + .584(\text{Acetic}) + .685(\text{H2S}) + 3.468(\text{Lactic})$$

Next we test the null hypothesis that $\beta_1 = \beta_2 = \beta_3 = 0$. Look for the table called "Omnibus Tests of Model Coefficients" shown in Table 16.3 below. The value is 16.334 and the p value is .001. Thus we reject H_0 and conclude that one or more of the variables are useful for predicting the odds that the cheese is

acceptable. Now, examine the coefficients for each variable and the tests that each of these is equal to 0.
The p values are .705, .090, and .191, indicating that the null hypothesis for each variable can be rejected.
 We conclude therefore that all three of the explanatory variables can be used to predict the odds that the
cheese is acceptable. Further analysis of these data is needed to determine which variables or
combinations of variables are needed to clarify the prediction of acceptability of taste in cheese.

Variables in the Equation

		B	S.E.	Wald	df	Sig.	Exp(B)	95.0% C.I. for EXP(B)	
								Lower	Upper
Step 1(a)	Acetic	.584	1.544	.143	1	.705	1.794	.087	37.004
	H2S	.685	.404	2.873	1	.090	1.983	.898	4.379
	Lactic	3.468	2.650	1.713	1	.191	32.086	.178	5777.335
	Constant	−14.260	8.287	2.961	1	.085	.000		

a Variable(s) entered on step 1: Acetic, H2S, Lactic.

Table 16.2

Omnibus Tests of Model Coefficients

		Chi-square	df	Sig.
Step 1	Step	16.334	3	.001
	Block	16.334	3	.001
	Model	16.334	3	.001

Table 16.3

Exercises For Chapter 16

1. Return to the data for Table 14.3 from your text.
   a. Complete a binary logistic analysis of the data using Group ID as the dependent variable and DRP as the explanatory variable.
   b. Write a summary of the outcome including significance tests and confidence intervals for the slope(s), the odds ratio, and the overall test of the null hypotheses.

2. Return to the data for Table 14.3 from your text.
   a. Complete a binary logistic analysis of the data using Group as the dependent variable and Retinol as the explanatory variable.
   b. Write a summary of the outcome including significance tests and confidence intervals for the slope(s), the odds ratio, and the overall test of the null hypotheses.

3. The following are data showing the relationship, for 64 infants, between gestational age of the infant (in weeks) at the time of birth and whether the infant was breast feeding at the time of release from hospital, "no" is coded as "0" and "yes" coded as "1."
   a. Organize the data in an SPSS spreadsheet. Use the WEIGHT BY option in SPSS as we did for Chi-Square analyses.
   b. Complete a binary logistic analysis of the data using Breast Feeding as the dependent variable and Gestational Age as the explanatory variable.
   c. Write a summary of the outcome for each analysis including significance tests and confidence intervals for the slope(s), the odds ratio, and the overall test of the null hypotheses.

Gestational Breast Feeding		
Age (Weeks)	Yes	No
28	4	2
29	3	2
30	2	7
31	2	7
32	4	16
33	1	14

4. Consider the data below on the effects of a particular treatment for arthritis. The response is Improvement; Sex and Treatment are the explanatory variables.
   a. Set up the data in an SPSS spreadsheet using numerical references for sex and treatment. (Note: You can ignore the total column).
   b. Code Improvement as 1 = some/marked and 0 = none.
   c. Use the WEIGHT BY option in SPSS to weight the data accordingly.
   d. Using Improvement as the dependent variable and Sex and Treatment as the explanatory variables, complete a multiple logistic regression.
   e. Write a summary of the outcome for each analysis including significance tests and confidence intervals for the slope(s), the odds ratio, and the overall test of the null hypotheses.

		Improvement		
Sex	Treatment	Some/Marked	None	Total
F	Active	6	21	27
F	Placebo	19	13	32
M	Active	7	7	14
M	Placebo	10	1	11

5. Shown below is a random sample of size 20 from the dataset "INDIVIDUALS" taken from the appendix of your text. Using this small subset of the data,
    a. Complete a binary logistic regression using Sex as the dependent variable and Earn as the explanatory variable;
    b. Complete a multiple logistic regression using Age, Education, and Earn as the explanatory variables.
    c. Write a summary of the outcome for each analysis including significance tests and confidence intervals for the slope(s), the odds ratio, and the overall test of the null hypotheses.

ID	Age	Educ	Sex	Earn	Job
499	25	4	1	53,883	5.0
878	25	3	1	32,000	5.0
1433	26	4	2	4,000	5.0
2344	26	1	1	11,440	5.0
2423	26	2	2	5,262	5.0
3324	27	4	2	18,380	5.0
5753	29	4	1	29,117	5.0
6651	29	3	1	25,000	5.0
7544	30	2	1	6,116	5.0
7588	30	3	1	25,000	5.0
7851	30	3	2	26,001	5.0
9258	31	3	1	45,000	5.0
9780	31	3	2	19,042	5.0
10273	31	3	1	32,017	5.0
11500	32	4	2	42,500	5.0
13590	33	4	2	7,370	5.0
13693	33	5	1	22,000	5.0
14783	34	3	1	82,500	5.0
14861	34	4	1	64,000	5.0
15907	35	4	1	18,000	5.0

6. Return to the data for Table 14.7 from your text.
    a. Complete a binary logistic analysis of the data using Group ID as the dependent variable and Decrease as the explanatory variable.
    b. Repeat the analysis using Begin as the explanatory variable.
    c. Repeat the analysis using End as the explanatory variable.
    d. Write a summary of the outcome including significance tests and confidence intervals for the slope(s), the odds ratio, and the overall tests of the null hypotheses.
    e. Is there anything here to be learned about using before and after scores versus using changed scores?

7. Shown on the next page is a random sample of size 24 from the dataset "MAJORS" taken from the appendix of your text. Using this small subset of the data,
    a. Complete a binary logistic regression using Sex as the dependent variable and GPA as the explanatory variable;
    b. Complete a multiple logistic regression using High School Math, Science, and English (HSM, HSS, HSE) as the explanatory variables.
    c. Complete a multiple logistic regression using high SAT Math and Verbal scores (SATM and SATV) as the explanatory variables.
    d. Write a summary of the outcome for each analysis including significance tests and confidence intervals for the slope(s), the odds ratio, and the overall test of the null hypotheses.

Sex	Major	SATM	SATV	HSM	HSS	HSE	GPA
1.00	1.00	760.00	500.00	10.00	10.00	9.00	3.80
1.00	1.00	640.00	670.00	9.00	6.00	5.00	2.00
1.00	1.00	520.00	360.00	9.00	8.00	7.00	2.73
1.00	1.00	620.00	480.00	9.00	9.00	9.00	2.87
1.00	1.00	670.00	500.00	7.00	7.00	8.00	2.21
1.00	1.00	447.00	320.00	9.00	10.00	8.00	1.92
1.00	2.00	630.00	500.00	8.00	7.00	8.00	3.50
1.00	2.00	670.00	440.00	9.00	7.00	6.00	2.96
1.00	2.00	770.00	540.00	10.00	10.00	10.00	2.97
1.00	2.00	620.00	570.00	10.00	10.00	10.00	2.81
1.00	3.00	610.00	390.00	9.00	6.00	6.00	1.84
1.00	3.00	470.00	330.00	6.00	5.00	6.00	2.39
1.00	3.00	515.00	285.00	5.00	7.00	7.00	.58
2.00	1.00	650.00	570.00	9.00	9.00	10.00	2.38
2.00	1.00	590.00	490.00	10.00	8.00	9.00	3.68
2.00	1.00	670.00	490.00	10.00	10.00	10.00	3.28
2.00	1.00	670.00	490.00	10.00	10.00	10.00	3.28
2.00	1.00	630.00	470.00	8.00	6.00	8.00	3.41
2.00	2.00	630.00	700.00	9.00	9.00	10.00	3.14
2.00	2.00	640.00	590.00	10.00	10.00	9.00	3.44
2.00	2.00	650.00	500.00	10.00	10.00	10.00	3.90
2.00	2.00	650.00	500.00	10.00	10.00	10.00	3.90
2.00	2.00	650.00	500.00	10.00	10.00	10.00	3.90
2.00	2.00	640.00	430.00	9.00	9.00	9.00	3.65

8. Suppose that we are working with some doctors on heart attack patients. The dependent variable is whether the patient has had a second heart attack within 1 year (yes = 1). We have two independent variables. One is whether the patient completed a treatment consisting of anger control practices (yes = 1). The other independent variable is a score on a trait anxiety scale (a higher score means more anxious). The data are shown on the next page.

a. Complete a binary logistic regression using Second Heart Attack as the dependent variable and Anger Treatment as the explanatory variable.

b. What are the odds of having a heart attack for the treatment group and the no treatment group?

c. Complete a binary logistic regression using Second Heart Attack as the dependent variable and Trait Anxiety as the explanatory variable.

d. Complete a multiple logistic regression using both anger treatment and trait anxiety as the explanatory variables. Specify that Anger Treatment is a categorical variable in the "Logistic Regression" window.

e. Write a summary of the outcome for each analysis including significance tests and confidence intervals for the slope(s), the odds ratio, and the overall test of the null hypotheses.

Person	HrtAtt#2	AngerTx	Trait Anx
1.00	1.00	1.00	70.00
2.00	1.00	1.00	80.00
3.00	1.00	1.00	50.00
4.00	1.00	.00	60.00
5.00	1.00	.00	40.00
6.00	1.00	.00	65.00
7.00	1.00	.00	75.00
8.00	1.00	.00	80.00
9.00	1.00	.00	70.00
10.00	1.00	.00	60.00
11.00	.00	1.00	65.00
12.00	.00	1.00	50.00
13.00	.00	1.00	45.00
14.00	.00	1.00	35.00
15.00	.00	1.00	40.00
16.00	.00	1.00	50.00
17.00	.00	.00	55.00
18.00	.00	.00	45.00
19.00	.00	.00	50.00
20.00	.00	.00	60.00

Chapter 17. Statistical Process Control

Topics covered in this chapter:

17.1 Process And Statistical Control

In this chapter we will apply statistical methods to a series of measurements made on a process in order to assess and improve quality. Quality in this context is defined as consistently meeting standards appropriate for a specific product or service. Using data is a key to improving quality, and statistical methods are applied to processes in service, manufacturing, and other organizations on a daily basis. Control charts are statistical tools that monitor a process and alert us when the process has been disturbed so that it is now out of control. This is a signal to find and correct the cause of the disturbance. This chapter emphasizes the use of systematic data collection and analysis, using charts, for making decisions about and for maintaining quality.   The goal is twofold: to make the process stable over time (manage the pattern of variability) and to make it within specifications.

x-Bar Charts For Process Monitoring

For charts of the mean of a process (*x*-Bar charts), we first choose a quantitative variable that is an important measure of quality. Consider a manufacturer of computer monitors who must control the tension on the mesh of fine vertical wires that lies behind the surface of the viewing screen. Too much tension will tear the mesh, and too little will allow wrinkles. Tension is measured by an electrical device with output readings in millivolts (mV). The manufacturing process has been stable with mean tension $\mu = 275$ mV and process standard deviation $\sigma = 43$ mV. This example is described in Example 17.4 in IPS.

The mean 275 mV and the common cause variation measured by the standard deviation 43 mV describe the stable state of the process. If these values are not satisfactory–for example, if there is too much variation among the monitors–the manufacturer must make some fundamental change in the process. This might involve buying new equipment or changing the alloy used in the wires of the mesh. We want to watch the process and maintain its current condition.

The operator makes four measurements every hour. Table 17.1 in IPS gives the last 20 samples. The first column of observations is from the first hour, the next column is from the second hour, and so on. There are a total of 80 observations. The table also gives the mean and standard deviations (stdev) for each sample.

To produce an *x*-Bar control chart using SPSS, follow these steps:

1. Click **Graphs,** and then click **Control.** The "Control Charts" window in Figure 17.1 appears.
2. By default, the "X-Bar, R, *s*" type of chart is selected.
3. By default, in the "Data Organization" box, "Cases are Units" is selected. Change this to "Cases are Subgroups".
4. Click **Define.** The "X-Bar, R, *s*: Cases are Subgroups" window in Figure 17.2 appears.
5. Highlight *obs1, obs2, obs3,* and *obs4,* then click ▶ to move them all into the "Samples:" box.
6. Click *sample,* then click ▶ to move *sample* into the "Subgroups Labeled by:" box.
7. In the "Charts" box, click on **X-Bar and standard deviation.**
8. Click "Statistics" and then click "Using within subgroup variation." See Figure 17.3.
9. Click on "Statistics" and the "X-Bar, R, s: Statistics" window appears as in Figure 17.4 following.
10. Enter the upper (339.5) and lower (210.5) confidence limits that you calculated based on earlier data collected for the process. See IPS for detailed calculation instructions.
11. Click **Continue.** Click **OK.**

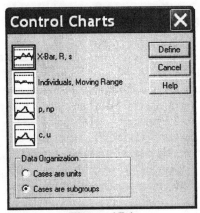

Figure 17.1

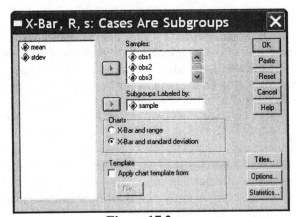

Figure 17.2

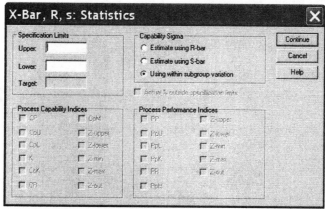

Figure 17.3

Figure 17.4 is the resulting SPSS for Windows output. The user specified UCL and LCL are indicated as "**U spec** = " and "**L spec** = ." In this *x*-Bar chart no points lie outside the control limits. Compare it to Figure 17.4 in IPS.

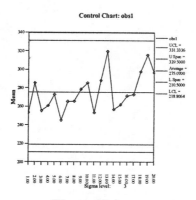

Figure 17.4

In practice, we must monitor both the process center, using an *x*-Bar chart, and the process spread, using a control chart for the sample standard deviation *s*. This is commonly done with a chart of standard deviation against time. If the process is out of control using this measure, it is important to eliminate this type of cause (*s* type cause) first and then return to monitoring the *x*-Bar chart.

To create an *s* chart, follow these steps:

1. Click **Graphs,** and then click **Control.**
2. Click on "Individuals, Moving Range."
3. Click "Define." The "Individuals, Moving Range" window in Figure 17.5 appears.
4. Highlight *stdev* then click ▶ to move it into the "Process Measurement:" box.
5. Click *sample,* then click ▶ to move *sample* into the "Subgroups Labeled by:" box.
6. In the "Charts" box, click on "Individuals."
7. Enter the value for the number of cases used in calculating the control limits, in this case, 4.
8. Click "Statistics" and then enter the user specified UCL and LCL (89.8 and 0, respectively) calculated using Table 17.2 and the instructions in the text. See Figure 17.6.
9. Click **Continue.** Click **OK.** The result is shown in Figure 17.7.

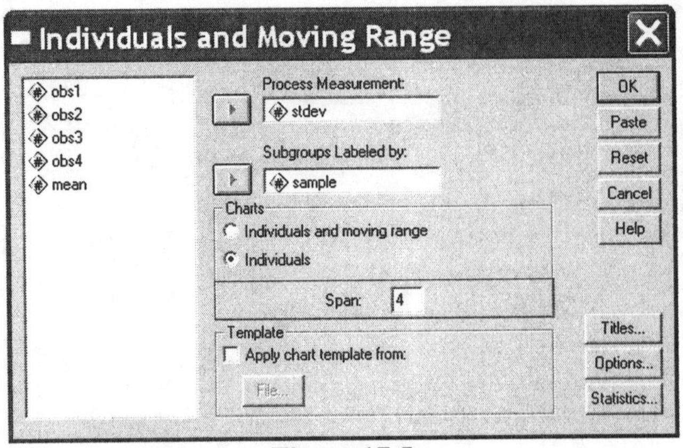

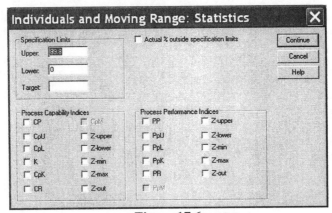

Figure 17.5

Figure 17.6

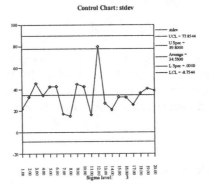

Figure 17.7

17.2 Using Control Charts

x-Bar And *R* Charts

Historically, the control chart for the mean was combined with the sample range rather than the standard deviation. To obtain an *R* chart using SPSS, follow the instructions given above except click on "*X*-Bar and range" in the "Charts" box (see Figure 17.2 above). Using the data from Table 17.1, we get the output shown in Figure 17.8. Such a chart is useful for pinpointing the "one point out" signal. Notice in the figure below that we have one high point but that the process returns to relative stability immediately after. Here we are using the sample range rather than the sample mean or standard deviation. Thus we have replaced the *s* chart with an *R* chart.

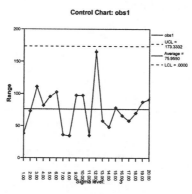

Figure 17.8

By reacting to single points that are outside of the control specifications often we may get false alarms. We can set our processes to monitor consecutive out-of-range signals, (six in a row, for example) and still maintain process control while limiting the number of false alarms. Generally process control systems are set to sound the alarm when a run reaches nine consecutive points either above or below the center line. On average, the runs signal responds more quickly than the one-point-out signal does.

Setting Up Process Control Charts

At the setup stage, control charts look back in an attempt to discover the present state of the process. Control charts for monitoring follow the process after the initial setup. By looking back at previous data, we are able to ascertain the usual mean and variability of the process and to use these values to calculate the values that we place in the "Specification Limits" boxes for the charts. For example, we can use the mean of the sample means to estimate μ. To estimate σ, we use the within sample variation. You can use SPSS, which uses sophisticated methods to calculate these estimates, or simply take the mean of all of the sample means and standard deviations as necessary.

To illustrate, open the data from Table 17.5 and create *s* and *x*-Bar charts. The results are shown on the left in Figure 17.9 following. This s chart shows us that two values of *s* were out of control. These occurred on Shifts 1 and 6. An analysis of the situations showed that the operator made an error on these two samples early in the shift and that no similar errors occurred later in the shift. Therefore we can remove these two samples and recalculate s from the remainder of the samples. The new s chart is shown to the right in Figure 17.9 for comparison purposes.

Now that the s type causes are removed, we can make an *x*-Bar chart using only the remaining samples. Figure 17.10 following shows this chart. The UCL and LCL have been specified as 49.862 and 47.081 as calculated using the directions in the text. Compare Figure 17.10 following to Figure 17.14 in IPS. We now see that, having removed within shift variability, there is considerable variability from shift to shift. Figure 17.10 below shows that three shifts are above the UCL and three shifts are below the LCL. We must now explore the cause of this shift-to-shift variability.

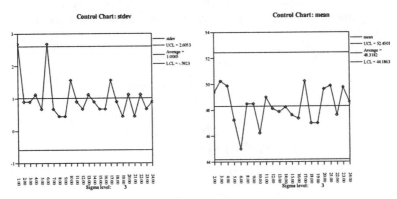

Figure 17.9

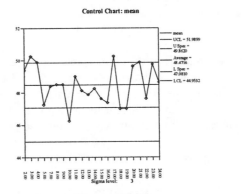

Figure 17.10

Comments On Statistical Control

This section of Chapter 17 of IPS offers comments and cautions about the practice of statistical control. These are reviewed briefly here. See the text for a full discussion. The first of these is that it is important to monitor the process and not restrict quality control to the final product. Thus choosing the "key points" at which to make measurements is critical. Secondly, it is important to consider how the samples are chosen. The variation within a sample should reflect only the item-to-item chance variation. Consecutive samples are appropriate when the stream of output is from a single person or machine. When the product contains output from several people or machines, the sampling process is more complicated and should be designed to reflect all persons and/or machines in a logical manner. Such statistical control is desirable because, if we keep the process in control, the final product will be within specified tolerances. Also, if changes are made to the production or service process, the effects of these changes will be immediately evident. See Examples 17.11 and 17.12 in IPS for a broader description of these issues.

Finally, we meet a discussion of control versus capability. Satisfactory quality is not the same thing as having a process that is in control. To ensure satisfactory quality, the product is compared to some external standard that is unrelated to the internal state of the process. Capability refers to the ability of a process to meet or exceed these external requirements placed on it. Example 17.13 in the text gives us an illustration. The purchaser of our computer monitors now requires new limits on the screen qualities of the monitors produced. While the production process is well in control, as it stands it does not have adequate capability to meet the new requirements.

17.3 Process Capability Indexes

SPSS provides measurements that can assist us in determining the capability of a service or production process. To explore these, return to the data from Table 17.5. First assess the percentage of the outcomes that meet certain specifications. To do so we create an x-Bar chart and specify the UCL as 50 (see Example 17.14 in IPS), the LCL as 40, and the target as 45. In addition, we can ask SPSS to print for us the "Actual % outside the specification limits." See Figure 17.11 below.

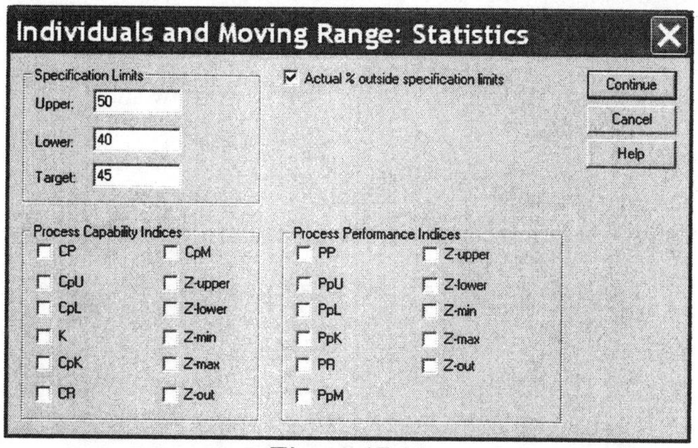

Figure 17.11

SPSS reports for us that 8.3% fall outside these process limits. While not exactly the figure given in IPS, it still means that 91.7% fall within the specified limits for viscosity. This is consistent with the output shown in the text.

Percent of processes meeting specifications, however, is a poor measure of capability. SPSS will report several other capability measures for us. Each gives us a numerical measure of process capability that permits us to monitor and strive for continuing improvement of a process.

Capability indexes use the fact that normal distributions are about six standard deviations wide. C_p and C_{pk} are two such measures. They are described in the text and illustrated in Figure 17.19 in IPS. Refer again to Figure 17.11 above. You will notice that SPSS will provide us with these Capability indexes. Larger values of both C_p and C_{pk} are better than lower ones.

If we return to the data from Table 17.5, we can obtain C_p and C_{pk} by clicking on the appropriate boxes in the "Individuals and Moving Range: Statistics" window (see Figure 17.11 above). The results are shown in Table 17.1 below. These values differ slightly from those given in the text, however, they give us the same overall information. The value of C_p is greater than one and this is quite satisfactory. The small value (less than one) of C_{pk} reflects the fact that the process center is not close to the center of the specifications.

Process Statistics

	Act. % Outside SL	8.3%
Capability Indices	CP(a)	1.264
	CpK(a)	.410

The normal distribution is assumed. LSL = 40 and USL = 50.

a The estimated capability sigma is based on the mean of the sample moving ranges.

Table 17.1

17.4 Control Charts For Sample Proportions

The most common alternative to x-Bar and R charts is the use of a p chart. A p chart is used when the data are proportions. To create a p chart, plot the sample proportions against the order in which the observations were taken. Example 17.20 of IPS uses p charts for discussing manufacturing and school absenteeism. Table 17.10 contains data on production workers and records the number and proportion absent from work each day during a period of four weeks after changes were made. To produce a p chart for these data, follow the steps outlined below.

1. Click **Graphs,** then click **Control.** The "Control Charts" window shown earlier in Figure 17.1 appears.
2. Click the icon for a "p, np" type of chart.
3. Click on **Cases are Subgroups.**
4. Click **Define.** The "p, np: Cases are Subgroups" window in Figure 17.12 appears.

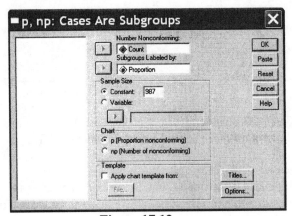

Figure 17.12

5. Click *Count,* then click ▸ to move *Count* into the "Number Nonconforming:" box.
6. Click *Proportion,* then click ▸ to move *Proportion* into the "Subgroups Labeled by:" box.
7. By default, in the "Sample Size" box, "Constant" is selected. In the empty box beside "Constant" enter the sample size. In this example, the sample size is 987.
8. Click **OK.**

Figure 17.13, on the following page, is the resulting SPSS for Windows output.

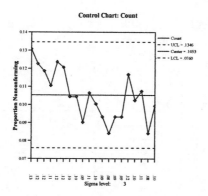

Figure 17.13

The *p* chart shows a clear downward trend in the daily proportion of workers who are absent. It appears that the actions taken to reduce the absenteeism rate were effective. The last two weeks' data suggest that the rate has stabilized.

Exercises For Chapter 17

1) The following samples are taken from a population with $\mu = 55.6$ and $\sigma = 12.7$.
   a. The table below contains 10 random samples of size 4 from this population.
   b. Create s and x-Bar charts from these data.
   c. What do your process control charts tell you about the process you are monitoring?

Sample	Obs1	Obs2	Obs3	Obs4	Mean	StDev
1	54.68	62.50	41.35	53.66	53.05	8.74
2	32.03	34.14	45.29	47.74	39.80	7.86
3	71.83	69.78	56.46	66.98	66.26	6.83
4	52.88	51.02	59.54	75.11	59.64	10.94
5	59.08	41.59	40.82	54.82	49.08	9.26
6	83.71	60.19	73.51	72.64	72.51	9.63
7	44.21	61.31	77.92	48.62	58.02	15.12
8	54.19	52.48	35.82	58.80	50.32	10.03
9	58.99	52.20	34.86	69.31	53.84	14.48
10	62.81	35.97	57.62	48.54	51.24	11.76

2) The registrar of a large university is interested in the grade distributions in large introductory level classes. At the end of the semester 10 final grades are chosen at random from each of 10 classes. The Registrar would prefer a distribution of grades with $\mu = 65$ and $\sigma = 15$.
   a. Create s and x-Bar charts from these data.
   b. Using the instructions in the text calculate and set the UCL and LCL.
   c. What do your process control charts tell you about the process you are monitoring?

Class	Obs1	Obs2	Obs3	Obs4	Obs5	Obs6	Obs7	Obs8	Obs9	Obs10	Mean	StDev
1	77.58	55.18	80.00	82.14	43.80	70.12	66.54	50.87	65.83	73.26	66.53	12.90
2	73.55	54.02	58.01	49.90	61.85	70.79	57.80	51.68	71.83	33.62	58.31	12.14
3	39.76	65.57	79.80	77.97	55.51	73.54	61.54	63.33	69.17	68.04	65.42	11.68
4	80.81	70.82	65.34	61.07	93.70	64.73	104.89	65.50	63.84	73.44	74.41	14.53
5	71.04	70.09	76.16	57.65	73.61	37.52	92.82	47.49	49.83	58.90	63.51	16.32
6	94.96	76.58	63.65	65.60	75.13	65.05	77.17	87.21	81.23	79.10	76.57	10.01
7	75.36	71.82	70.15	47.95	60.22	55.23	49.81	68.70	51.96	74.64	62.58	10.74
8	85.79	36.44	70.98	64.57	61.77	74.82	66.46	76.45	85.35	48.51	67.11	15.51
9	70.09	86.72	67.36	61.82	71.37	61.53	41.97	63.80	100.46	59.88	68.50	15.84
10	57.78	75.64	55.85	67.23	50.84	72.59	80.91	94.08	39.83	66.95	66.17	15.76

3) A manufacturer of lead for mechanical pencils has introduced a new machining process. To monitor the new process, 15 samples of size 10 are taken during the first shift after installation. The leads are built to a specification of a mean of .3 mm with a variability of .10. The data are on the following page.
   a. Calculate the UCL and LCL for the lead production.
   b. Create s and x-Bar charts from these data showing the user-specified UCL and LCL.
   c. Using the instructions in the text calculate and set the UCL and LCL.
   d. What do your process control charts tell you about the process you are monitoring?

Sample	Obs1	Obs2	Obs3	Obs4	Obs5	Obs6	Obs7	Obs8	Obs9	Obs10	Mean	StDev
1	1.44	1.12	2.30	.65	.10	1.72	.17	.85	.02	.76	.91	.75
2	1.01	.34	.66	1.78	.02	.12	.81	1.04	.36	1.94	.81	.65
3	.31	−.92	.93	−.47	.83	.80	.78	.26	.22	.90	.36	.63
4	.24	.46	1.27	.45	.86	.76	.14	1.56	.66	1.65	.80	.53
5	.01	.20	.81	.48	1.04	.72	.46	1.10	.10	.21	.51	.39
6	1.08	.80	.29	.72	−.34	.87	.58	.14	.82	-.34	.46	.50
7	.47	−.09	.99	.35	.29	.77	1.33	.71	.22	.37	.54	.41
8	.72	.75	−.15	.74	.32	.02	.95	.72	.41	.33	.48	.36
9	.66	.59	.39	.46	.19	.63	.47	.61	.51	.26	.48	.16
10	.57	.61	.72	.51	.45	.33	.22	.42	.57	.54	.49	.15
11	.43	.46	.54	.58	.31	.90	.54	.50	.41	.52	.52	.15
12	.32	.46	.49	.61	.50	.40	.21	.56	.50	.29	.43	.13
13	.41	.66	.18	.43	.65	.54	.52	.37	.74	.58	.51	.17
14	.37	.57	.63	.45	.45	.47	.41	.34	.54	.14	.44	.14
15	.40	.33	.71	.36	.58	.52	.40	.66	.42	.39	.48	.13

4) A basketball coach working with six-year old children is interested in how the players are doing with their free throws. The coach sets up a system in which all free throws are monitored by an assistant and logged. The coach then randomly selects 10 free throws from 15 consecutive practices and scores them for the proportion completed and the number of misses. The data are shown below.
   a.   Create a p chart from these data.
   b.   What does your process control chart tell you about the process you are monitoring?

Sample	Obs1	Obs2	Obs3	Obs4	Obs5	Obs6	Obs7	Obs8	Obs9	Obs10	Proportion	Count
1.00	.00	.00	.00	1.00	1.00	1.00	1.00	.00	1.00	1.00	.60	4.00
2.00	1.00	1.00	1.00	1.00	1.00	.00	1.00	.00	1.00	1.00	.80	2.00
3.00	1.00	1.00	1.00	1.00	.00	1.00	.00	1.00	1.00	.00	.70	3.00
4.00	1.00	.00	1.00	1.00	1.00	1.00	.00	1.00	1.00	1.00	.80	2.00
5.00	1.00	1.00	.00	1.00	1.00	1.00	1.00	1.00	.00	1.00	.80	2.00
6.00	1.00	.00	.00	1.00	.00	.00	.00	1.00	1.00	.00	.40	6.00
7.00	1.00	.00	.00	1.00	1.00	1.00	1.00	.00	1.00	.00	.60	4.00
8.00	.00	.00	.00	1.00	1.00	1.00	.00	1.00	.00	1.00	.50	5.00
9.00	1.00	.00	.00	1.00	.00	1.00	.00	.00	1.00	.00	.40	6.00
10.00	.00	1.00	.00	.00	1.00	.00	1.00	1.00	.00	.00	.40	6.00
11.00	1.00	1.00	.00	.00	.00	1.00	1.00	.00	1.00	1.00	.60	4.00
12.00	1.00	1.00	1.00	.00	1.00	.00	1.00	1.00	1.00	1.00	.80	2.00

5) Return to Exercises 1 through 4 and calculate the actual percent outside specification limits and report both C_p and C_{pk} for each data set.
   a.   For each situation, what do these values tell you?

6) Eight samples of size 4 are taken from a population with $\mu = 275$ and $\sigma = 43$. The data are shown on the next page.
   a.   Create an s chart from these data.
   b.   Using the instructions in IPS, calculate the UCL and LCL for the s chart.
   c.   Remove any samples that exceed the UCL and LCL.
   d.   Create the x-Bar chart from these data.
   e.   What do your process control charts tell you about the process you are monitoring?

Sample	Obs1	Obs2	Obs3	Obs4	Mean	StDev
1	308.88	286.77	404.53	259.71	314.97	63.00
2	253.67	253.65	273.09	223.46	250.97	20.50
3	470.01	232.84	138.33	355.02	299.05	144.42
4	198.14	268.50	227.12	280.13	243.47	37.83
5	280.82	279.77	283.92	255.95	275.12	12.90
6	303.09	297.51	306.67	293.12	300.10	5.99
7	273.42	281.25	277.30	275.53	276.87	3.32
8	282.12	279.75	276.70	264.45	275.76	7.86

7) Experienced dart players rarely miss a throw. The data are shown below are for the first four throws of each of 18 successive trials by the same player.

   a. Create a p chart from these data.

   b. What does your process control chart tell you about the process you are monitoring?

Sample	Obs1	Obs2	Obs3	Obs4	Proportion	Count
1	1.00	1.00	.00	1.00	75.00	1
2	1.00	1.00	.00	1.00	75.00	1
3	1.00	1.00	1.00	1.00	100.00	0
4	1.00	1.00	1.00	1.00	100.00	0
5	1.00	1.00	1.00	1.00	100.00	0
6	1.00	.00	1.00	1.00	75.00	1
7	.00	.00	1.00	.00	25.00	3
8	1.00	1.00	1.00	1.00	100.00	0
9	1.00	1.00	1.00	1.00	100.00	0
10	1.00	.00	1.00	1.00	75.00	1
11	1.00	1.00	1.00	.00	75.00	1
12	1.00	.00	1.00	1.00	75.00	1
13	1.00	1.00	1.00	1.00	100.00	0
14	1.00	1.00	1.00	1.00	100.00	0
15	1.00	1.00	1.00	1.00	100.00	0
16	1.00	1.00	.00	.00	50.00	2
17	1.00	.00	1.00	.00	50.00	2
18	1.00	1.00	.00	.00	50.00	2

Index